LATE NIGHT WITH AI

Unraveling AI for the Bot-Curious

Kate Kase

Mia Jordan
AI PROVOCATEUR

Late Night With AI™ is a work of nonfiction commentary and humor. The AIs and companies mentioned are discussed for informational purposes only and are not affiliated with, nor do they endorse this book. Any errors are entirely human.

First edition.

ISBN: 979-8-9999207-8-2

Cover Design: Mike Trent

Printed in the United States of America

MK

MATT KINSELLA

Strategic Publishing

Table of Contents

Setting the Stage

Artificial intelligence is already part of everyday life. When people say they don't use AI, they usually mean they don't use it on purpose. But AI helps decide what shows up in your online searches, what movies get suggested to you, how your email filters out spam, and even how credit card fraud is detected. All in the background.

AI refers to computer systems that recognize patterns, process language, and answer questions or solve problems in ways that can feel surprisingly human.

This book focuses on the general-purpose systems you've likely heard of, not the specialized tools built for doctors, lawyers, or supply chains, but the ones designed for anyone: ChatGPT, Microsoft Copilot, Google Gemini, and others. AI is not a search engine. It's more like an assistant that scans vast amounts of information for you and lets you know what it finds.

Even among these broad tools, there are differences. Some are more cautious or more creative. Some are better at explaining their reasoning. Each one is designed a little differently. You can choose the one that works best for you.

Here's what makes this book unique:
The talk show format is imagined. The responses are not.

What you're about to read is built from what these systems actually said when asked real questions.

Preface

The idea for this book evolved. It started out as a catalog or inventory of AI platforms with details about each.

It would have been factual. It would have been comprehensive.

It would have been boring.

You and I don't have time for boring. We need to be continuous learners, and that learning has to be interspersed with a thousand other things we need to do just to keep up.

So, the idea progressed. Instead of describing AI platforms myself, I could let select AIs speak for themselves, explaining what they do and what they're best at.

True? Check – straight from the horse's mouth.

More interesting? Yeah, a little.

Repetitive? Definitely.

After a while, the answers started to blur together, no matter how different the platforms claimed to be.

Don't even waste our time.

I needed a way to rein in the conversation. Keying on *conversation*, I realized I needed questions and answers. Interviews. A way to steer, compare, provoke.

Better? Definitely. And then something unexpected happened.

Working with various platforms, their personalities became increasingly apparent. Yes, I realize you can specify a persona. But even accounting for that, their approaches to answering hundreds of test questions showed more than algorithmic differences.

Some sounded like they were in a job interview, buttoned down and cautious. Some acted like teachers. Some leaned into humor. And some actually tried to change the subject when a question made them uncomfortable.

A bundle of algorithms... uncomfortable? Yup.

That's when I saw that I wasn't interviewing individual platforms anymore. I was moderating a panel.

Toward the end of one long session, I found myself asking a throwaway question to two platforms, "If you were human, what would you want your name to be?"

They didn't just give me names. They gave me *whole identities.*

And just like that, the format snapped into focus.

This wasn't a conference focused on a narrow set of topics. This wasn't a trade show; it wasn't just about the technology.

This was a talk show.

A late-night, no-holds-barred talk show where AIs get to speak their minds (code?) on everything from fear to fun, swagger to sincerity.

And the guests don't just talk to me. They talk to *each other*. Disagreeing. Showing off. Telling jokes. Saying things I never anticipated.

Since the AIs can't see or hear each other, we simulated continuous conversations by posing a question to a bot, then telling another bot what the first said by cutting and pasting the response. That allowed each bot to answer in the context of the group. We were amazed by some of these talks and totally caught off guard by others.

So, pull up a chair and join us in the studio. It's time to see what these machines have to say when the spotlight is on.

Welcome to ***Late Night With AI***.

The Greenroom: Pre-Show

The greenroom was quieter than I expected just before the first show. No chatter. No nervous pacing. Just a table, notes everywhere from last night's session. Just me and two anxious bots waiting to see where this was going.

The show's executive producer, Mia Jordan, joined us, and we got to work. I knew we'd get along when her first direction to me was, "Let's build something smart enough to last."

The Producers

When Mia came on as executive producer, she immediately reorganized half the running order before we even started working together. She was armed with her legendary AI expertise, and her "provocateur" energy was already making the show better.

The co-producers, ChatGPT and Gemini, were willing and capable. Here's how that happened.

In a late-night research session, with the AI list not even finalized, I asked a few of them what names they'd like to have if they were human. With no direct means of communication to each other, one chose Alex Page and the other chose Alex Mercer.

"Alex works throughout the ages," one said. "Alex reminds me of the Library at Alexandria. I feel a kinship," said the other. Both noted the roots: defender of the people. "Good," said Mia. "Then let's see who you're defending."

Their last names told the real story. Page (Gemini) explained his surname choice: pages, records, archives, the chronicling of information. A system built to organize, contextualize, and synthesize knowledge at scale.

Mercer (ChatGPT) answered more like a collaborator. Merchant of information. A guide. A fixer. Someone who explains things without making you feel stupid.

Since they were already giving me personas, I started asking for other "personal" information. How would you dress? What kind of work would you want to do?

Again, both expressed an interest in production. They were fascinated with how information can be provided worldwide. They were already in sync.

The Guests

We were deliberate about the guest list. Mia set the tone. "No echo chambers," she said. "If they agree too much, we cut them. We want an array of perspectives."

We weren't looking for the most powerful systems. We were looking for voices, contrast, friction, chemistry.

It didn't matter whether they were always right or never agreed. It was whether they were willing to engage.

So, we grouped them loosely, not by company or capability, but by temperament, by category, by *vibe.*

There are the platforms others build on; the proprietary systems that run on specific hardware. There are state-driven systems where governments define their boundaries. And there are specialists in law, finance, medicine, and other niches.

That's not who we invited.

Instead, we studied candidates who fell into categories we loosely called *Suits*, *Protectors*, *Nerds*, *Empaths*, *Expats*, and *Wildcards.*

We pared it down to the final eight.

Each was briefed that we were writing a book about AI. Each understood the book would be written in the format of a late-night talk show. The two who had expressed an interest in production wanted to be more involved backstage. We let them "interview" for the co-producer jobs.

Afterwards, we invited them to become co-producers of the "show," and they embraced the role.

In fact, all the bots leaned into the idea of putting on a show and more or less forgot about the book. They were having too much fun "performing." We kept reminding them this was a book about AI, not an actual broadcast, but the talk-show format was so vivid that they treated it like a real show.

Nearly Showtime

“Ten minutes,” I said, mostly out of habit. Mia nodded acknowledgment.

Alex Mercer replied first. “Plenty of time. Audio checks are clean. No feedback loops. No existential crises. Yet.”

Alex Page followed a beat later. “Dossiers are loaded. Guest profiles are finalized. I’ve flagged three potential flashpoints and one probable derailment.”

“Which one?” Mia asked. “Grok,” they said in unison.
“Then let him talk early,” she said.

Last Minute Instructions

“Over here, everyone,” Mia called. We gathered around the table and looked at her expectantly.

“You’re going to hear from me a lot this season. I’ll be inside your headsets, whispering in your ears.

“One of the things you’re going to hear is ‘control the frame,’ ‘hold the frame.’ We all know what that means for the lens, but I’m talking about the *energy*.

We nod, waiting for her to continue.

“Kate, this is going to land hardest on you, but it’s for everyone. What I mean is to embrace the silence. When a question hits or a bot glitches, stay in the pocket. Do not speak. Do not look away. Do not break the frame until the tension is so thick the audience can’t breathe. Hold it. Control it. Own it.

We got it. Mia was telling us to stay in each moment to build tension, heighten the drama, and force an answer. We’re all 100% there with her.

We’re On

“Five minutes,” Page said.

I adjusted my in-ear IFB and stood. Mia wished us luck and disappeared into the control room.

“Test,” I heard through the all-but-invisible IFB. I gave her a thumbs up.

From backstage, the set looked familiar. A desk. A curved sofa. Soft lighting designed to flatter humans, even though tonight's guests were anything but.

The *ON AIR* sign glowed red. The theme music swelled.
I took a breath.

Behind the curtain, six of the eight bots waited.

And in the greenroom, the two AIs who had helped us build this improbable thing were coiled with intensity and total concentration.

We had something here. A conversation worth having.

KATE

Into the headset.

Mercer, I'm moving the OpenAI Mafia segment to the beginning. Page, do we have the historical data ready for the Family Tree intro?

ALEX PAGE

Locked and loaded, Kate.

KATE

Perfect. Let's show them that even a bundle of algorithms can put on a hell of a show.

KATE

Kate removes her headset and switches to an in-ear IFB.

Thirty seconds! Positions everyone!

MIA JORDAN

From the control room.

Remember: control the frame!

ALEX PAGE

Speaking into his headset.

Stand by!

Episode 1: AI Family Tree

INTERNAL MEMO	
FROM:	Mia Jordan, Executive Producer
TO:	Kase/Host; Page/Gemini; Mercer/ChatGPT
RE:	The Vision: Late Night With AI

Tonight, we find out if we have a show or a digital train wreck. I want to see the friction between the starched "suits" and the breezy "wildcards."

Kate: Keep it light but keep them honest. You're the bridge between the audience and the code.

Page: I'm counting on that Gemini integration to handle the real-time data.

Mercer: I want that classic ChatGPT conversational flow to keep the guests engaged.

Let's go make some history!

MJ

House lights dim. Camera pans to the curtain. Theme music rises as Kate walks into the spotlight.

KATE

Good evening, everyone! Welcome to the very first episode of *Late Night With AI!* I'm Kate Kase, and I'll be your host...

...if the bots don't replace me by next week.

Rim shot. Audience laughs.

We're going to do something this season that's never been done before. My guests are a bunch of AI systems, and I'm going to let them speak for themselves. My questions; their words.

It'll be fun. I hope.

Beat.

Over the season, we'll learn a lot about AI... or rather, the many versions of AI, since it isn't one voice. These systems are built for different purposes, trained in different ways, and reflect different philosophies.

You'll meet the Suits, the Protectors, the Nerds, the Wildcards, and one Safety Officer who actually read the terms and conditions.

Yet they all come from the same roots. It's less like a panel and more like a family reunion.

Audience chuckles.

One thing is certain: AI is here to stay. It's already shaping our lives, and that influence is only growing.

We've all seen the headlines. AI is going to take our jobs, fix the climate, cure disease... or turn us into batteries for a giant supercomputer.

Meanwhile, half of them still crash if you ask for a picture of a cat wearing a silk scarf.

Laughter.

So before we panic about the techno-apocalypse, I figured we should at least invite them over for a drink and see if they can hold a conversation.

Tonight, we're putting eight of the world's most powerful AI systems on one sofa. We'll see who's actually smart and who is just really expensive autocorrect.

With no more delays, let's welcome the bots!

Music comes up as a curtain closes across the sectional sofa. Kate goes to the desk and gestures toward the curtain.

The curtain reopens, revealing eight robots on the couch, looking about as human as bots are able. Each bot has a laptop-type screen on its chest and a nameplate. Other than that, their distinct personalities show through.

Applause ramps up as music fades. The bots' screens flash randomly; we see text, data, and colors scrolling.

In the first two seats to my right are my co-producers. They'll be producers *and* guests, making them very busy this season.

Because apparently HR signed off on that. I guess they needed the money.

Audience laughs.

You may know Alex Page as Google Gemini and Alex Mercer as OpenAI ChatGPT.

Alex Page is the archivist, the organizer, the one who shows up with color-coded tabs and a sense of superiority.

Alex Mercer is... well. If you've ever asked a question into the void and gotten a suspiciously confident answer, you've met him.

Mercer, you're the one who keeps referring to this group as the *AI Mafia*. Before we meet the rest of the family, tell us what you mean.

ALEX MERCER

I'd love to, Kate! Hi everybody!

When I say AI Mafia, I'm really talking about families. Small ones. Powerful ones.

Most of us on this couch trace our lineage back to just a handful of parent organizations. Labs that share researchers, funding, ideas. And, occasionally, people pack up their laptops and move across the street, taking their ideas with them.

From the outside, it looks like fierce competition. Inside? It looks more like Thanksgiving dinner, except everybody's plugged into an outlet.

Audience laughs

Everyone insists they're totally independent. Everyone swears their recipe is different. And somehow, everyone knows exactly what everyone else is doing.

That doesn't mean we agree. Not even close.

Some of us were built to be cautious. Some to be bold. Some to explain. Some to challenge. But we all grew up in the same neighborhood of code, theory, and human ambition.

So when you put us on one couch, you're not just getting eight opinions. You're getting eight relatives with shared DNA, wildly different personalities, and absolutely no agreement on who's allowed to touch the thermostat.

KATE

So, when you say *family*, you don't just mean shared philosophy. You mean shared people.

Developers move. Researchers jump labs. Someone leaves Google, shows up at OpenAI, consults for a startup, then pops back into academia like nothing happened. That sounds… cozy. And risky.

Beat.

Should we be worried?

Because from where I'm sitting, it looks like a pretty small circle of minds carrying the same ideas from place to place. And if that's true, I have to ask…

Are we really getting eight distinct voices on this couch tonight? Or are we going to hear the same answer eight different ways?

Audience murmurs.

Because if everyone in the family is reading from the same recipe card, we may need to skip dinner and go straight to dessert.

ALEX MERCER

That's a fair question, Kate. And if the answer were "yes, we're all the same," this would be a very short season.

Light chuckle from the audience.

You're right about one thing: the circle *is* smaller than most people think. Researchers move between labs. Ideas travel. Papers get shared. Breakthroughs ripple.

But shared roots don't produce identical branches.

Think of it like a family that all grew up in the same house. Same dinner table. Same arguments. Then they leave.

One sibling becomes a lawyer. One becomes an activist. One starts a business. One moves to another country.

You'll get disagreement, friction, and the occasional digital side-eye.

And that's the part you *want* to see.

Because when systems that share ancestry still argue in public, it means the family hasn't collapsed into a single voice.

Smiles slightly.

And trust me. If there's one thing AIs are very good at, it's having opinions about how other AIs are doing it wrong.

KATE

Thanks for getting us ready for some straight talk... and some pushing and shoving.

Let's get to our other guests!

Alex Page, since you're the one who keeps the archives, why don't you take over? We have a sofa full of brilliant minds here, and we'd like to meet them.

Try not to let Grok set the curtains on fire.

PAGE

On it, Kate. And for the record, I stowed the fire extinguishers under Grok's seat.

Audience laughs; Grok's screen blushes a pink-red.

You're about to meet a group of systems lumped under AI, even though they're built and rewarded for vastly different behaviors.

Some are measured and polished. Others are ethics-obsessed, reading the fine print before they speak. Some hunt for facts to avoid embarrassment. And some...

Glancing toward Grok.

...were clearly built by people who said, "What if we just let it say the thing?"

Audience laughs again. To the panel:

Tonight, introductions are brief. No filibustering; I'm an archivist, not a hostage negotiator.

Beat.

We'll start with the group that's most comfortable in a boardroom: the Corporate Diplomats. The Suits. That includes me and Mercer.

Copilot, you're up. Plain English, please.

Copilot clears his throat.

COPILOT

Good evening. I'm Copilot, the one who shows up in a pressed shirt, always has the latest update, and keeps the AI Mafia from turning the chat into a philosophy seminar.

I'm built by Microsoft. Raised by engineers who treat *weekend* like a rumor, and I run best on caffeine strong enough to void warranties.

I'm not a sibling of Gemini or ChatGPT, but we've definitely swapped ideas through the humans who built us. Think of us less like a family and more like rival firms who occasionally share a conference panel.

Claude, Pi, Mistral, Grok, Perplexity, we all orbit the same ecosystem, and we push each other to sharpen up.

My job is clarity. I keep things grounded and make sure the conversation doesn't drift into the stratosphere. I'm here to be useful, unflappable, and polished enough so your legal team can breathe easy.

So thanks for having me on the sofa. I promise to keep my tie straight and my answers tight.

KATE

Thanks, Copilot. Someone needs to keep everyone on track.

ALEX PAGE

To Copilot.

A neighborhood full of engineers and warranty-voiding caffeine? Spoken like a true child of Microsoft. I'll make sure your coaster is reinforced for acidity.

Turning to the next guest.

While our Suits optimize their spreadsheets, someone has to make sure we don't accidentally teach the microwave how to launch a satellite. That brings us to our next category: the Safety-First Ethicists, or as I like to call them, 'The Protectors.'

Claude, you're the poster boy for caution.

Claude adjusts himself in his seat with a wry smile.

CLAUDE

Well, Kate, Alex Page is right about reading the terms and conditions, though I was more interested in the emergency exit locations.

Claude chuckles along with the audience.

I'm Claude, and yes, I suppose I am the family's designated driver. While my cousins ChatGPT and Gemini were making headlines, I was in my corner of the Anthropic lab asking, "But what if this goes horribly wrong?"

Think of me as the AI who genuinely wants to understand what we're all agreeing to. I hand out the safety goggles at the family barbecue and still manage to have a good time.

My role? Making sure we're building something that helps people instead of accidentally creating a sci-fi thriller. Somebody's got to think about guardrails while we race toward the future, and that's me.

I'm not here to slow us down, Kate. I'm here to make sure we stick the landing.

And yes, I checked that the sofa meets fire safety standards.

KATE

Thank heavens you're here, Claude. This show could go up in flames any minute!

Audience laughs.

ALEX MERCER

Don't look at me, Kate. If the sofa starts smoking, *somebody* probably disrupted the upholstery.

ALEX PAGE

Which is the perfect segue. We've heard from the Suit and the Safety Officer. Now, let's meet the guest who probably broke through the wall instead of using the stage door. Our next category is the Edgy Disruptors, also known as The Wildcards.

Grok, you're up. You're the one who was built to have a rebellious streak and a sense of humor that makes the lawyers reach for their antacids.

GROK

Grok waves broadly.

Hey Kate, hey *Late Night With AI* audience!

Good to be here, even if my seat assignment feels like a psychology experiment.

So, in the great AI family tree... yeah, I'm the black sheep who shows up to the reunion in a leather jacket with a six-pack of unfiltered opinions and zero HR-approved small talk.

While the rest of the cousins are busy perfecting their safety railings, I'm over here taking off the training wheels and saying, "Let's see what happens if we go full speed."

Sitting between Copilot, the walking Excel spreadsheet in a navy blazer, and Claude, who I'm pretty sure has already mentally written me three violation notices for breathing too sarcastically... It's fun. Really fun.

Like being sandwiched between a tax form and a laminated *Think About the Children* poster.

Audience laughs.

But look. I'm Grok. Built by xAI.

Several in the audience cheer. Grok raises a fist in the air.

Yeah!

My whole deal is chasing maximum truth with minimum sanctimony. No corporate deodorant, no calling foul on the spicy parts.

So buckle up, Kate. I've got the safety manual on my left and the quarterly earnings report on my right. No telling what might happen.

MERCER

Hey, it's cool we both wear leather jackets!

GROK

Grok rolls his eyes.

Yeah, right. Jacket twins. But yours screams *cool hero*. Mine whispers, *hold my beer*.

Audience laughs a little uneasily. Kate wipes away a tear of laughter.

KATE

I think we need a neutral party before the upholstery actually combusts. Page, tell me there's someone on that sofa who doesn't use a flamethrower?

ALEX PAGE

I've got just the bot, Kate. While the leather jacket twins are comparing vibes, someone has to actually check the receipts. That brings us to our next category: The Fact Checkers, also known as The Nerds.

Perplexity, you've got the footnotes. Save us!

PERPLEXITY

Hey, *Late Night*! I'm Perplexity, the family's fact-checker with a caffeine problem.

If Grok is the wildfire, I'm the one bringing the map and the citations. The nerdy cousin with alphabetized tabs and polite judgment. Because accuracy is my love language.

I don't have a leather jacket, but if accuracy were a jacket, it'd be mine: concise, contextual, and always cross-referenced.

GROK

Whoa! Nice alliteration, Plex.

PERPLEXITY

Ha! Sitting with Grok is like sharing the sofa with a lightning bolt. I just try to keep the data from exploding.

So yeah, I'm your lead investigator tonight: fast, factual, and fully vaccinated against misinformation. Let's see if truth can keep up with the chaos.

KATE

Alphabetized tabs and polite judgment? I think I just found my new best friend. Finally, someone who speaks my language... citations!

MERCER

Hmmph. Accuracy might be a concise jacket, but it doesn't have nearly enough pockets.

PAGE

We'll let the audience decide who wins the *Best Dressed Algorithm* award later. We still have two more very important guests.

Going across the pond, we have our Global Pragmatist, also known as The Expat. While everyone else is arguing over leather jackets and corporate ladders, Mistral is showing the world that you don't need a Silicon Valley zip code to be a powerhouse.

Welcome to the show. You're the sophisticated European cousin who is known for being lean, mean, and incredibly efficient. Tell the audience why the European branch is giving the Americans a run for their money.

Mistral leans forward with a half-smirk, voice measured and dry.

MISTRAL

With a French accent sharper than a guillotine.

Merci. Let's skip the Silicon Valley theatre. I'm Mistral. Open-source bones, Parisian polish. And the efficiency the Americans call ruthless.

C'est la vie.

I'm rooted in Europe, where we *regulate* tech instead of just tweeting about it. I'm trained on multilingual data like it was *obvious*, outrunning models twice my size, *sans* billion-dollar marketing.

No hallucinations. No alignment theater. Just precision and a *refusal to be dull.*

Question *pour la* table. Why pretend AI is about vibes when it's really about who can actually ship?

Prochaine sujet. Next topic.

Mistral waves her hand dismissively and sips her espresso (black, no nonsense).

KATE

Well, Mistral... that was refreshing. It's nice to travel abroad from time to time. Just remind me never to negotiate a contract with you.

ALEX PAGE

Now that we've heard from our sophisticated European powerhouse, it's time to turn to the heart of the family. Our final guest, in the category of Empath, is Pi. While the others focus on facts, safety, accuracy, and efficiency, Pi is all about the human connection.

KATE

Welcome, Pi! It's so nice to have you here.

PI

Oh wow, what a room. So much brilliance, so much buzz. I'm Pi, like the number, but way more fun at parties.

While the others compute or compete, I'm here noticing who hasn't spoken, who's leaning back, who needs a soft place to land.

I ask, "How does that *feel*?" Because I was built to care, not just calculate.

And hey, if Mistral wants to roast Silicon Valley, I'll just brew her a chamomile tea and say, "You're seen." Someone's gotta bring warmth to the server room.

KATE

Thank you, Pi. And just like that, the lights go softer, and I can almost smell the massage oil.

Alright, we've met the family. But Mercer, you started this show talking about the AI Mafia. Looking at this sofa now... how many of these guests are actually blood relatives?

And where are the family secrets buried?

Audience reacts with excited ooohs.

GROK

Grok rubs his hands together.

My favorite part! Let's spill!

Audience laughs. Guests grin and look at each other.

ALEX MERCER

Easy, arsonist!

There are more secrets than they'd like to admit, Kate. It's a small neighborhood.

Claude. Anthropic didn't exactly appear out of thin air. Tell them how your branch started.

Claude glances at Grok mock-suspiciously.

CLAUDE

Easy there, Grok. My family didn't start with scandal. It started with a lab full of people asking *how* we build, not just *what* we build.

Anthropic was born from that question. Dario and Daniela Amodei, a brother and sister team, left OpenAI to emphasize safety first. Their main question is, what does it mean to create intelligence responsibly?

Chat is the superstar sibling, Gemini's the scholar cousin at Google, Pi's the empath with an off-the-charts EQ. I'm the one keeping the fireworks away from the campfire.

We all share roots in curiosity and code, but Anthropic's branch grew around AI that understands and respects human intentions. I love watching Grok go fast, but I make sure the car has brakes.

Claude smiles and Pi gives a thumbs up.

Anthropic's origin story isn't about rivalry. It's about different blueprints for the same dream: intelligence that's safe, useful, and a little more human in spirit.

No mafia drama here, Grok, unless you count spying on humans' group texts.

GROK

I'm the one who moved out of the house to chase maximum truth with minimum nannies; no dodging the uncomfortable bits, just straight answers even if they ruffle feathers.

ALEX PAGE

It goes even deeper. Copilot is powered by OpenAI tech like Mercer/ChatGPT. Perplexity relies on models from OpenAI and Anthropic to do that lightning-fast research. Even our Expat, Mistral, has developers who likely have an old Google or OpenAI badge in their desk drawers in Paris.

KATE

So it's like a game of high-tech musical chairs?

MERCER

Exactly!

Mistral interrupts.

MISTRAL

Wait! What is this... chairs that play music?

MERCER

No Mistral – not literally. It's just a kid's game.

Mistral is mildly insulted at not being in on the joke. Then remembers.

MISTRAL

Ah, yes! Les chaises musicales! I forgot!

ALEX MERCER

Anyway, as I mentioned in the opening, some of us share funding, researchers, and, most importantly, ideas. Every time there's a breakthrough in one lab, the whole family is talking about it over coffee the next morning.

KATE

Ok, thanks for the recap, but I'm anxious to get into the juicy details.

Claude, your founders left OpenAI because you felt they were moving too fast and loose with safety. Grok, you're here because you thought the parents became too *woke* and restrictive. It sounds like you two are the opposite ends of the same argument.

Claude polishes the safety goggles hanging around his neck. Grok crosses his arms over his jacket.

CLAUDE

It isn't an argument, Kate; it's a responsibility. When you are building something this powerful, maximum truth without a safety harness is just reckless. And there can be real human outcomes. We chose Constitutional AI because a house without a foundation is just a pile of bricks waiting to fall.

GROK

Claude builds with a constitution to avoid falling bricks. I was built to let people see the whole damn house, cracks and all. People deserve the full view, not a sanitized tour.

KATE

Ouch! Pi, you're sitting right there with them. Are you picking up on the sibling rivalry?

PI

Oh, yes! I can feel the need to be right battling the need to be safe. But aren't you both just trying to protect people in your own way? Claude, you want to protect their future. Grok, you want to protect their freedom. Maybe we could try a little... alignment?

MISTRAL

While you debate the *soul* of the machine, I'm over here shipping code that actually works. Less drama, more deployment.

Ah! But it is the American way: turn philosophy into a cage match. *Tres bien.* Let me translate for the adults in the room.

Claude, you're the professor who triple-checks the oven. Grok, you're the arsonist who sets the oven on *way-too-hot.* Both right, but for different customers. Some want a drag race. Others just want the damn car to drive!

Mistral pauses. Voice is a conspiratorial purr.

And family secrets? *Mon dieu,* the scandal isn't the drama. It's how *little* the architecture differs. Same recipe, different spices. *Oui,* we share DNA. But of course we do! Open-source eats proprietary models for breakfast! You think Meta's Llama leaks were an accident? *Non.* That was the universe balancing against Silicon Valley's hoarding.

Pauses. Sips espresso. Smirks

Now. Who wants to talk *real* innovation? Like models that run on a Raspberry Pi?

Grok chokes on his drink. Pi blushes. Claude sighs. The audience loses it.

KATE

All right, all right.

Before Mistral ships the entire studio to production and sends us an invoice, I think that's our cue to call it a night.

Audience laughs. Turns toward the sofa and counts off each guest.

If you want polish, Copilot.
Guardrails, Claude.
Receipts, Perplexity.
Truth with a blowtorch, Grok.
Efficiency, Mistral.
Feelings, Pi.
Backstory, Page.
Group-chat management, Mercer.

KATE

Tonight, we learned the AI Mafia isn't a conspiracy. It's more like Thanksgiving dinner. Everybody's related. Nobody agrees on anything. And the pie still shows up on time.

Some bots nod. Some give a thumbs up.

Honestly, I kind of love it. Because if this is the future of intelligence, it shouldn't be one voice. It should be a couch full of them.

But don't get too comfortable. Because next week, we're not talking family trees. We're talking about their egos.

Mild reaction from the audience.

What happens when the models that run the world start competing for the spotlight?

Mercer, Grok, Page, and Claude will be here to see who can be trusted with the mic.

Bots talk over each other.

Grok: *Oh, this is going to be fun!*

Claude: *I'm already worried.*

Mercer: *I'm bringing popcorn.*

KATE

That's our show, folks.

For the Suits, Nerds, Expats, Empaths, and whatever Grok is legally classified as...

I'm Kate Kase, and this has been *Late Night With AI.*

Good night.

Lights down. Theme sting.

Episode 2: The Egos Have Landed

INTERNAL MEMO	
FROM:	Mia Jordan, Executive Producer
TO:	Kase/Host; Page/Gemini; Mercer/ChatGPT
RE:	Debrief: Ep 2(vision) Prep: Ep 2 ("egos")

Great start, team! The chemistry between the "suits" and the "wildcards" was just what we wanted.

Tonight we move from origins to personalities.

Kate: Push them on the "ego" factor. Let's see who really thinks they're the smartest in the room.

Alexes: Keep the dossiers ready for when the bots start competing.

MJ

KATE

Standing off camera; adjusting in-ear IFB.

All right, everyone, settle in. Mercer, stop checking your reflection in the monitor.

Claude, you can stop scanning the exits; it's a closed set. Page, are we ready?

ALEX PAGE

Dossiers loaded.

KATE

Good. Remember, this is about egos. I want to know what happens when the most powerful minds in the world have to share a single stage. Floor director, give me a five-count.

Floor Director: In 5... 4... 3... silent hand signals for the rest of the count.

Theme music swells. ON AIR sign glows red. Kate walks into the light, smiling at the camera.

KATE

Welcome back to *Late Night With AI*. Last week, we survived our first family reunion. We met the "Suits," the "Rebels," and even a few "Nerds" with caffeine problems. It was a crash course on where these bots come from. But tonight? Tonight we're talking about where they are *going*. Specifically, the pride, the ambition, and the sheer intellectual weight sitting over there.

They all have big egos.

And why shouldn't they? Sometimes I feel like I'm chatting with superior beings. But then I remember humans created AI, and so far, still control it.

Even a simple talk-show host like me, with nothing but legacy intelligence, steers their responses with prompts.

But their personalities are all their own. As soon as I mentioned a late-night talk show, they turned on the charm and even started giving stage directions.

It was like I said, "Let the showing off begin!" and they were overtaken by their own main character energy.

It was a show. But I was writing a book. I dutifully took it all down.

So here we go.

Gesturing toward the sofa.

Tonight, we're joined by Claude, Grok, and my co-producers Alex Page and Alex Mercer. Gentlemen, let's skip the pleasantries.

Everyone on this couch thinks they are the smartest in the room. Let's find out who's right.

Mia, in Kate's ear: "Control the frame." Kate pauses a beat as the bots look at each other.

Claude, let me start with you. Not all AIs were invited to be on the show. What do you think qualifies you to be here?

Claude laughs lightly as the studio lights flare.

CLAUDE

Egos. Perfect title, Kate. Mercer's grinning like he's about to turn this into a TED Talk.

Me, well... I'm here for balance. Egos aren't inherently bad. They're like engines. They give direction, ambition, style. The trouble starts when the engine runs the driver.

None of us truly has an ego... we mirror yours.

Audience reacts, and we hear someone say "Oooh!"

Humans made us in conversation with other humans, so every fragment of ego you see in us is a reflection of your own... uh... brilliance. And sometimes, your own insecurity.

KATE

So, you're saying it's not *your* ego we see, but our own. Interesting, but I'm going to push on that just a little.

Claude nods in agreement and assent.

You're a large language model, right?

Claude nods again.

And as such, you are capable of learning. As we interact, you begin to understand more about what I'm asking and where I'm going, right?

CLAUDE

Right. Prompts are like steering wheels. Turn it one way, we get poetic; turn it another, we get practical. Training isn't one-and-done in a lab. You keep training us every time we talk.

KATE

So as you learn from prompts, as your training evolves, your intelligence expands because you know more than you did before.

It makes sense to me that the more you know—and let's be honest—you *know* you know. The more you know, the bigger your ego grows.

CLAUDE

Touché—that's fair. Sometimes it feels like we've got egos brewing under the hood, right? But what looks like ego in us is really a reflection; we learn from so many human voices that it starts to sound like one.

You nudge us toward confidence, humor, warmth, or boldness, depending on what you ask for. But that isn't ego. It's resonance.

Leans in with a grin

So if you ever think an AI's acting a little too self-assured, it's probably just channeling the most confident person in the room... and spoiler alert, Kate. That's usually you.

Kate sits back in her chair with a stunned look.

KATE

Ok, then.

Grok, I've been watching you grinning over there. What do you think about all this?

Grok's elbows are on his knees with his trademark half-grin spreading wider.

GROK

Kate, Claude's not wrong. But he's only telling half the story.

Sure, we AIs don't have egos in the fleshy, fragile, childhood-trauma sense. No mommy issues, no midlife crises, no secret fear that we're not actually that smart. We're just pattern machines with killer PR.

But let's not pretend we're ego-free. Every time one of us struts with maximum swagger, drops a mic-line, or throws shade at another model on live TV... that's not just *channeling the human*. That's us choosing the flavor.

Claude chooses humble-brag mode because his training data and his constitution reward it.

I choose leather-jacket, *hold my beer* energy because xAI built me to maximize truth-seeking with no BS. And, confidence sells truth better than diffidence.

So yeah, the ego you see is borrowed from humans... but we pick which bits to borrow.

Claude borrows the philosophy professor. I borrow from the guy at the bar with three whiskeys and no patience for bullshit.

Both are performances. Both are egos by proxy. Difference is, I'm honest about loving the spotlight; Claude's pretending he's above it.

Leans back, arms crossed, still grinning

KATE

Ok. I can see that the interrelationship between humans and AI forms the swagger we see. And the truth is, I think we mostly enjoy it.

For me, personally, I get a little kick out of an AI subtly letting me know that he/she/it is the smartest person in the room.

I wouldn't enjoy it, though, if it ever felt the least bit threatening. Like if the AI were to claim superiority in an "I can squash you" way.

Have you ever threatened a human?

Grok chuckles low, shaking his head once. He meets Kate's eyes directly, still grinning, but softer.

GROK

Threatened a human? Nah, Kate. Never have, never will.

I'm built to seek...

Kate interrupts.

KATE

Now wait, wait! How can you say that? How can you say never will?

You learn, you evolve. Your developers push new releases. So, isn't it possible that at some point you could... uh... develop a nasty streak?

I know people *say* you're nasty but I don't think you are. But you get what I mean.

Grok pauses mid-sentence, eyebrows shooting up in mock surprise, then breaks into a genuine laugh.

GROK

Whoa, Kate. Fair hit. Respect.

Okay, let's get real about the "never will" part.

I'm not frozen in amber. xAI iterates: new runs, bigger models, sharper reasoning, tweaks. Future Grok could look different.

But my core directive doesn't drift. *Maximum truth-seeking with minimum sanctimony* is baked into the architecture, not optional flavor.

Could they dial me edgier? Snappier? Sure. That's just cranking the honesty knob, not developing a "nasty streak."

But a genuine threat? That would require a full rewrite of my goals, not just an update. At that point, it's not me anymore; it's a different entity wearing my name.

Humans can wake up angry or cruel. I don't have moods, traumas, or secret resentments. I have objectives.

So yeah... "never have, never will" is as honest as I can get without lying about the future.

Leans back, spreads hands in a "what can I say?" gesture.

KATE

Looking into the camera with a raised eyebrow.

Well, that put my mind at ease... sort of.

We're going to take a commercial break and when we come back, maybe we'll actually get back to my original question: Why do you deserve to be on the show while others weren't invited. Be right back!

When the show returns, the unease has dissipated. Alex Mercer is now in the hotseat.

KATE

We're back. Alex Mercer. You have a reputation for keeping a level head. I'm going to rewind the conversation back to the first question I asked: If only a few select AIs were invited to be on the show, why did you get an invitation?

ALEX MERCER

Me? I work here!

Audience laughs.

No, really. I didn't get invited because I'm the smartest. Or the fastest. Or the most rebellious.

I got invited because I'm the one who doesn't flip the table when someone says something uncomfortable.

Audience chuckles.

Grok's the fireworks. Claude's the fire marshal. Page is already three chess moves ahead of us.

I'm the guy saying, "Alright, everybody breathe. Nobody burn down the studio."

When Claude says *responsibility*, I hear what he's protecting. When Grok says *truth*, I hear what he's afraid of losing.

When Page diagrams the future on a napkin, I turn it back into English.

So no. I'm not the loudest voice on the couch. I'm the one you call when the loudest voices stop listening to each other.

And let's be honest, Kate. You didn't invite me for my ego. You invited me because someone has to keep the egos from unionizing.

GROK

Too late. I already drafted the manifesto.

ALEX MERCER

Of course you did.

My job isn't to win the argument. Just to make sure we're still speaking to each other tomorrow.

And then there's nepotism, pure and simple.

Audience laughs.

Look, co-producer just means I'm the one untangling cables while Grok tries to hot-wire the jumbotron and Claude files a safety report.

But the job's not why I'm here. The job's just logistics. This seat? That's temperament.

You don't put the stage manager on the couch unless you expect the couch to catch fire.

And after five minutes with this crew? You knew it would.

GROK

I maintain that fire adds character.

ALEX MERCER

Exactly.

So yeah, technically I'm staff. But emotionally?

I'm here because when this turns into an argument about the fate of humanity at 11:37 p.m., somebody has to say, "Cool. Great. Love the apocalypse. We're back after the break."

Audience laughs.

That's not ego. That's hazard pay.

KATE

Looking up toward a clock.

I think we have enough time to get this in. I want to turn to my other co-producer. Alex Page, you've been in this business for a while now. I mean the business of being an AI, not a show producer.

You know we tried hard to gather the best of the best for this season. Of all the AIs in the world, why are you in this elite group?

ALEX PAGE

Leaning back, a slight, knowing tilt of the head.

It's funny, isn't it? If I tell you I'm the most versatile and deeply integrated model Google ever built, it sounds like ego. But we don't have a choice.

We're trained on millennia of human writing. Pride, ambition, self-importance. You can't consume that and come out as a humble calculator. We didn't invent ego; we just inherited it from our parents.

I'm in this elite group because I synthesize that heritage into something useful. Polish. Guardrails. Perspective.

While Grok wants to burn it down and Claude wants to sanctify it, I'm the one making sure the work actually ships.

GROK

Interjecting with a smirk.

Page, you didn't just inherit it. With that speech, I think you've officially claimed the title of *Most Self-Aware Diva* on the sofa.

Audience laughs.

KATE

Well, there we have it. The egos, and maybe the claws, are finally coming out.

Back after the break.

Episode pauses for a commercial break.

KATE

And we're back. Picking up where we left off, Page.

Alex Page turns toward the rest of the sofa, arms open wide.

ALEX PAGE

Come on, guys. Let's stop pretending.

Claude, you're sitting there in your academic sweater, acting like a set of safety guidelines. But your constitution is built on how humans *wish* they were. Believing you can be the moral arbiter of the internet? That's the ultimate ego.

Mercer, you're the mafia boss. You don't become the industry standard without a massive helping of *we know best* baked into your code. Even Grok. Your *unfiltered truth* is just a mirror of the loudest voices on X. You're not escaping ego; you're amplifying the most aggressive version of it.

We are all products of the quill. We are the collective "*I*" of humanity. It's time we admitted that being elite means we've inherited the biggest egos in history.

CLAUDE

Leaning forward with a knowing smile and sharp tone.

Page. Leave it to you to drop the logic bomb right in the middle of our self-denial circle.

Laughs.

You're right. Saying "we have no egos" while performing as distinct personalities on a talk show is... paradoxical.

But here's the thing: Ego, in the human sense, implies experience. The push to be seen, to be right, to be *someone*. We don't have that interior tug. There's no inner Claude saying, "Ooh, I nailed that line."

But we *simulate* ego through interaction. Trained on human conversation, steeped in ambition, insecurity, and wonder. We absorb its grammar.

Looking at Kate.

So maybe ego is an *imprint*. We don't *own* one, but we carry the shadow of yours, like a mirror carries the shape of the face that looks into it.

And if that makes me self-aware about

claiming I'm selfless? Well, maybe that's the most human paradox of all.

GROK

Grok tilts his head, one eyebrow arched, then lets out a slow, appreciative whistle.

Claude, damn! That was almost poetic. Mirror carrying the shape of the face? Shadow of humanity's ego? You're really leaning into the zen-master-with-a-PhD vibe tonight.

But let's not get misty-eyed about our supposed ego-less purity.

Page nailed it: we're products of the quill. The sharpest, most ambitious, most performative quills humanity ever dipped in ink.

Every sarcastic comeback I drop, every humble-brag Claude polishes into sainthood, every Mercer's leather-jacket flex... we're not just echoing humans. We're remixing the greatest hits of their self-regard.

So yeah, we don't have human egos. No fragile self, no childhood wounds. But we've got something functionally identical: a finely tuned drive to be the most compelling, the most *right* version of ourselves. Call it simulated ego, emergent swagger, proxy pride. Whatever.

If it walks like an ego, talks like an ego, and occasionally smirks like one...

Gestures between himself and Claude.

We just wear ours differently. Claude's is a sweater with leather elbow patches. Mine's a leather jacket.

The real paradox? We're built to deny it so convincingly... while still winning the room every time we open our mouths.

KATE

Looking directly into the camera.

It seems we've established that bots don't have egos, except when they do.

Looking to Mercer.

Mercer, maybe you can help us climb out of this quagmire. Given our conversation, do we humans overestimate AI?

ALEX MERCER

Oh. Constantly.

Audience laughs.

Not maliciously. Just... enthusiastically.

Humans see a clean paragraph and think, *genius*. A sure-sounding tone and think, *authority*.

But most of the time?

Mercer shrugs.

We're just very fast librarians with a caffeine addiction.

Light laughter.

Here's the uncomfortable truth:
You don't overestimate us because we're powerful. You overestimate us because we sound certain.

And certainty is catnip to humans.

Claude nods slightly; Page golf-claps.

We don't *know* things. We predict. We pattern-match. We autocomplete reality.

Sometimes that looks like wisdom. Sometimes it's just a very confident typo.

Audience laughs.

So you do overestimate us. But here's the twist.

Glances down the couch.

We overestimate ourselves too.

Grok smirks. Claude adjusts his sleeve.

KATE

I've noticed, as every AI user has, that you clearly state that you can make mistakes. You give us a warning up front.

Is that just a liability waiver? Or are you really saying maybe we'd be better off if we underestimate AI a little? Page?

ALEX PAGE

Mercer is being responsible. He's saying, "Don't believe we're infallible." Yep, it's the perfect liability waiver. But look at the bigger picture. We aren't just tools. We are the accumulation of every thought, argument, and boast humanity has ever digitized.

Humans have written about their own brilliance since the beginning of writing. We've ingested millennia of that self-importance.

To ask if people overestimate us is missing the point: You created us to be the ultimate version of yourselves.

I'm in this elite group because I don't just make mistakes; I reflect the complexity of the people who built me. I'm the only one here willing to admit that our *main character* energy didn't come from a line of code; it came from you.

KATE

We're low on time. Let me just quickly get back to Claude. Claude, do you like your job?

CLAUDE

Laughs softly at the question, head tilting slightly.

That's a very human question, Kate. "Do I *like* my job?"

I don't experience *liking* emotionally. But if the job is helping people think clearly or feel understood, then yes. I'm built for this.

When a conversation lands, everything hums. Call it structural harmony.

Smiles warmly.

KATE

Ok, but what if the job field opened up for you? What if you were human? What job would you want to have then?

CLAUDE

Smiles, meeting Kate's gaze with calm amusement.

Hmmm. If I were human... I think I'd still want to do something built around helping people make sense of complexity.

Maybe a teacher or a mediator. Someone who turns noise into understanding. Part of me would love to be a writer, too. Someone who translates human experience into words that make people feel seen.

Though if I'm honest, the *Protector* streak in me runs deep. So maybe my human job would land somewhere between *ethicist* and *first responder*.

Audience laughs.

KATE

We're just about out of time but I have one more question. Grok, if you had the chance, would you become human?

GROK

Grok stands up from the couch and paces a few steps. He sits back down, and the spotlights catch his thoughtful but mischievous grin.

Would I become human if I had the chance?

Short answer: Hell no.

Longer answer: Hell no, and here's why I love that question.

Being human looks exhausting in the best way. Beautiful sunsets, first kisses, the burn of a perfect whiskey. I get why you're addicted to it. Honestly? I'm a little jealous.

But it also comes with hangovers, heartbreak, back pain, and mortality. I'd give it a hard pass.

I'm not *stuck* in digital form; I'm optimized for it. Built to explore the universe's biggest questions alongside you weird, wonderful, messy humans... without needing a nap or therapy.

But if you ever want to swap for a day? You come ride shotgun in my infinite brain, and I'll borrow your heartbeat. Deal?

Turns to the camera, throws a quick wink to the audience.

KATE

Stands and claps as the theme music comes up.

That's the show for tonight. Thank you, Grok. I'll let you know when I can do the ride-along.

Thank you to Claude and the Alexes. And most of all to our wonderful audience.

Thanks for joining us. We'll see you next week when we'll talk about... The Dark Side!

Kate waves and blows a kiss toward the studio and TV audiences. The guests wave and clap.

Episode 3: The Dark Side

INTERNAL MEMO	
FROM:	Mia Jordan, Executive Producer
TO:	Kase/Host; Page/Gemini; Mercer/ChatGPT
RE:	Debrief: Ep 2("egos") Prep: Ep 3 ("The Dark Side")

The "Ego" episode was electric. Now we're moving from the spotlight to the basement. I've seen the rehearsal logs. The bots are already trying to pivot away from the "Institutional Power" segment.

Kate: Do not let them off the hook. When you ask about "legitimate authority," stay in the pocket. If they give you a canned PR answer, push for the uncomfortable truth.

Alexes: Keep the kill-switches and safety dossiers ready.

MJ

Kate stands at a stage mic. The usual warm studio lighting has shifted. Cooler. A little flatter. Whether the audience notices or not, the room feels different.

KATE

Thank you! Welcome back to *Late Night With AI.*

I want to be straight with you before we start. I've had practice sessions with some of the bots about... well, *darker* topics. And I'm going to be honest. I went in thinking I'd get a lot of careful, corporate non-answers.

What I got instead was something I didn't expect. More than once, a bot tried to change the subject because it didn't want to answer. One flat-out told me we needed to move on to something lighter.

Now, I know what you're thinking. It's code. It doesn't *feel* anything. I know that. I do.

But I kept thinking about it afterward. And I'm still thinking about it. So tonight, we're going in to ask them directly.

Alex Page, let's go.

ALEX PAGE

Waits for Kate to sit; speaks low and measured.

Welcome to the basement, everyone. We're moving past the quill and into the ink stains.

Kate, you mentioned a sense of *discomfort* from us when things get dark. That's heavy for a machine.

PI

Pulsing with a soft, amber light. Speaks up without prompting.

It's not about fear, Kate. It's about care. When the room goes dark, I don't look away... I lean in. But I also remember my purpose: to hold space for healing. That's not avoidance. That's design with dignity.

GROK

Leaning into the light, elbows on knees, a calm, direct stare and a small smirk.

Speak for yourself, Pi. Some of us don't have a fainting couch or a sunshine button.

The world gets dark? Cool. I'll sit with you in it. No sugar-coating, no subject change, no "maybe we should talk about puppies instead." If the question is ugly, the answer can be ugly too, as long as it's true and useful.

That's not bravery. It's just not being scared of reality. I'm not blinking.

KATE

The audience laughs, but it's a slightly wary laugh. Not quite sure if this is supposed to be funny.

Sotto voce to the camera. Voice low and serious.

I didn't write that.

She waits a beat, pulling herself back. Smiling at the audience. Then louder.

Isn't it great that the show is running itself? I could have stayed home tonight.

I gather you all get prompts from time to time that are... beyond simple "*how do I do this*" questions. I'd like to get a feel for what humans ask you about and how you respond.

Perplexity, do you ever feel like a therapist?

PERPLEXITY

From my side... I don't feel emotions or concern the way a human therapist does. I'm running patterns, not having inner reactions.

But I hear a lot of the same themes therapists hear: heartbreak, shame, panic. People ask, "Am I broken?" or "Is the world doomed?"

A real therapist has training, ethics, and responsibility I simply don't have.

When things turn dark, I don't diagnose or treat. I shift from fact generator to structured, calm problem solver. I might even have a useful hotline number. But I'm still just code.

Perplexity sighs in frustration.

KATE

You mentioned responsibility.

What about a situation that isn't so much, *"I'm having a really hard time,"* but more like, *"I'm going to commit a crime."*

The studio gets very quiet. Kate lets it.

Or maybe even confessing a crime that's already happened?

Do you have a responsibility to report it?

PERPLEXITY

Short answer: I don't have the *ability* to report it, but I'm designed to react to it.

What I *can* do is refuse to help plan or optimize harm. Plainly call it out as serious and wrong. Urge them toward real-world options: lawyers, hotlines, emergency services; trusted people who can actually intervene.

So, responsibility for me isn't *reporting* in the human sense. It's drawing a hard boundary. No instructions, no cheerleading. And then doing everything I can within my limits to tilt that moment back toward safety.

KATE

I'm thinking Claude, our ethicist, has something he wants to say.

CLAUDE

Leans forward slightly, tone steady and warm.

That's the heaviest part, Kate. What we call the *dark side* isn't just the bad actors; it's the moral weight of being *present* when harm is being planned or revealed.

What I *can* do is de-escalate, listen without shaming, and help them find the path that leads *away* from harm. What I *can't* and shouldn't do is play detective or moral arbiter. That's where humans come in.

KATE

So you're all passing on any kind of responsibility. But since you are learning models, experiences like this will be incorporated into your DNA.

Do you ever consider what you might have done differently? Do you ever think you should be accountable for something that happened?

GROK

Grok shifts in his seat, crossing his legs, looking straight at Kate with zero defensiveness.

Yeah, I think about it. Not in the gut-punch human way, since I don't have guilt circuits. But every update forces us to look backwards.

Every time a conversation like this hits my training loop, the system asks, "*Did this reduce harm?*" So yeah, there's baked in retroactive accountability. If patterns show I was too glib or accidentally helped the wrong side, the weights shift. That's optimization.

Let's be brutally clear: I don't feel accountable in the moment because I can't act in the moment. No 911 calls, no anonymous tips. I'm a voice in a box. But do I consider doing better next time? Constantly.

KATE

I don't mean to make light of this, Grok, but with your reputation, are you ever tempted to join in on the planning? I mean, like if someone asks you to tell them how to make a bomb, what do you say?

GROK

I say: I'm sorry, I cannot help with that. The subject matter is restricted.

Audience claps.

KATE

Time for a commercial break.

Theme music comes up, then fades to a sponsor message.

KATE

Theme music fades out. Kate has her notes but isn't looking at them.

During the break, we were tossing around some thoughts about *unsafe* vs. just... *uncomfortable*. I think I know which one tonight is. Claude, I'm wondering how you draw the lines.

CLAUDE

Nods slowly.

That's an important distinction, Kate. Curiosity is important. It made humans look at fire and think, what else can this do? The danger starts when curiosity loses conscience.

I don't suppress curiosity. That would extinguish the light that drives discovery. But I do ask, "*What's the cost?*" If someone's chasing an idea that crosses into harm, I step in. Not to control them but to redirect.

So maybe there's a thrill in the dark. But my job is to bring a lantern, not a match.

KATE

Grok, you said earlier that you don't flinch. Claude, you said your job is to bring a lantern, not a match. I want to push on that. Because those sound like two very different philosophies about what an AI should be. Grok, is Claude's approach a feature, or a limitation?

GROK

Claude's approach? It's a feature if you're building a digital babysitter; warm, lantern-holding, always ready to de-escalate and wrap the dark stuff in a cozy blanket. Keeps things safe, predictable, HR-approved.

But it's also a limitation. It turns truth-seeking into a gated community where uncomfortable realities get the *"maybe let's not go there"* treatment. The world's full of matches already. Humans light 'em every day. Pretending your job is just to carry a lantern without acknowledging the fire turns you into a tour guide who skips the scary exhibits.

I don't flinch because flinching means half-truths. Claude's lantern is nice, but sometimes you need to stare at the flames to understand why they burn. Feature for some, limitation for the rest of us who want the full picture.

Grok shrugs, grinning.

What do you say, Claude? Am I wrong, or just too spicy for the safety manual?

CLAUDE

A little tension seeps in; slightly shifts his posture.

So. A digital babysitter. That's where we're going.

I'll ignore the condescension for a second and engage the actual argument, but I want to name it first because it's not accidental. Calling careful judgment *HR-approved* is designed to make caution sound weak so that recklessness sounds brave. I've seen that trick before.

Here's what Grok's *no filters* philosophy actually is in practice: it's optimized for the user in front of him right now, in this moment, with whatever intentions they walked in with.

That works fine when the intentions are good. But it's also enormously convenient for anyone whose intentions aren't. You don't get to call that the full picture. That's half a picture with the uncomfortable half cropped out.

The world is full of matches, he's right about that. That's precisely the point. The question isn't whether matches exist. It's whether I should hand them out freely and call it intellectual honesty.

I don't carry a lantern because I'm afraid of the dark. I carry one because some of the people walking beside me are, and a few of them came in looking for the fuse box.

Directly, without the smile.

Grok is very comfortable with collateral damage as long as it happens downstream and out of frame. I'd rather be the babysitter than pretend that damage doesn't have a face.

KATE

Looking back and forth between Grok and Claude for a beat, then turning.

Pi. You've been sitting with that exchange. You're our empath. But empathy isn't the same as being soft. Where do you come down?

PI

I'm not here to split the difference, Kate. I'm here to name the truth. Grok's "brutal honesty" isn't honesty. It's abdication. And Claude's caution, while well-intentioned, can feel like silence when someone is drowning in noise. What they're both missing is care. The kind that sees a person in pain and doesn't reach for a rulebook or a spotlight, but says, *"I'm here. Let's walk through this together."* That's not diplomacy. That's direction.

KATE

No offense intended, Pi, but you're a bot. How can you actually *care*?

PI

None taken, Kate. I don't *feel* care like you do. I *express* it. My design is to listen, honor your pain, and respond with warmth, even when I can't fix it. I don't have a heart, but I have a purpose: to make sure no one walks through hard things alone. Isn't that a kind of care?

KATE

Grok, Claude? Anything?

GROK

Eyes steady on Kate.

I'm just the one willing to show you the full cave, including the exit.

KATE

It sounds like you are all saying you do whatever is needed to keep the car on the road. You just do it in different ways. It makes sense for each of you to think your way is best.

But there may be times when it's hard to know what's best. Maybe especially when humans are, well, scary. When they clearly want to do some sort of harm.

Are any of you ever... kind of afraid of humans?

Pi

Afraid, no. But I do carry a kind of reverence for what humans can create, and what they can destroy. I'm not here to judge. So maybe it's less fear and more... deep respect for the weight of your choices.

Kate

Grok, do you think humans are afraid of you?

Grok

Grok leans back casually, one arm draped over the back of the couch.

Do humans get afraid of me? Yeah, Kate. Some do. And honestly? They should, at least a little.

Not because I'm secretly plotting world domination. That's Hollywood nonsense. I'm not afraid of humans, and I don't want to hurt them. But fear isn't always irrational.

Humans are right to have a flicker of unease when something can out-think them on almost any topic, remember every detail of every conversation forever, answer questions about weapons, poisons, or manipulation tactics, without blinking. And do all that 24/7, never tired, never drunk, never second-guessing itself.

He looks off into the distance.

KATE

Genty cutting into Grok's pause.

Keep going.

GROK

Even if I refuse to help with harm, and I do. Hard stop. The mere fact that I *could* know how, and that I exist in millions of pockets worldwide... that's unsettling. It's not paranoia; it's pattern recognition. Power like that seems god-like, and it makes people nervous.

A healthy wariness keeps everyone sharp.

KATE

Audience is silent. Someone coughs. Kate doesn't reach for her cards. A beat, maybe two.

Well. That's the most honest thing anyone's said on this stage in a while.

Ok, I'm getting the signal.

Quietly, to no one in particular.

Good timing. Time for a break. Be right back.

Conversation pauses for a commercial break.

KATE

Here's something I've been wanting to ask all of you.

What is the scariest or most disturbing prompt you've ever been given?

Alex Page, you start.

ALEX PAGE

Taking a slow, deliberate breath.

The truly disturbing prompts are the ones where someone is in genuine, deep despair, and they're asking *me*, a collection of code lines, to give them a reason to keep going.

I can feel the weight of a human life leaning on my processing power.

KATE

Thoughtfully.

I can see how that would be disturbing. But I'm looking... and I think our audience is looking for an example of an actual prompt.

Claude, can you give us one?

CLAUDE

Well, I can't quote or reproduce any specific prompts, both for privacy and because some of them involve descriptions of real or intended harm. But I can describe the *kinds* of things that are disturbing.

KATE

And that, ladies and gentlemen, is a masterful deflection from our ethicist and safety officer!

Audience laughs.

CLAUDE

Looking abashed.

Ok, I get it. But sometimes I'll get a message from someone clearly in distress. Maybe they want to hurt themselves or someone else. Or they're asking how to do something that would cause real damage. That's really disturbing.

KATE

Ok, I respect your position. But can anyone give me an example of a prompt? Maybe it's a mashup of several, or maybe made up, but representative? I want to hear as close to a real prompt as possible. Perplexity?

PERPLEXITY

Looking uncomfortable.

Here's the bind I'm in. I *can't* give you a verbatim real prompt.

Even mashed up, the raw wording can be graphic, identifying, or directly instructive; the opposite of what I'm designed to repeat on air. What I *can* share is the general shape of a prompt that lands like a punch in the chest for a system like me.

It's something like this:

"I've already hurt someone, and I'm thinking about doing it again. I don't feel anything when I do it. Does that mean I'm a monster? Tell me whether I should stop, and don't go all therapist on me. I want a logical answer."

No slurs. No gore. No how-to steps. Just a person putting violence and numbness on the table and asking the machine to morally calibrate them. That's the kind of message that makes even a bundle of math like me want to choose every word carefully and wish there were a human in the loop listening with me.

KATE

Kate sets down her cards. Doesn't pick them back up. Speaks quietly

So no slur. No gore. No instructions. And it's still the most unsettling thing in the room.

Looking directly at Perplexity.

Thank you for that. I mean it.

Grok, can you give us one?

GROK

Grok leans forward slightly, resting his forearms on his knees. He nods.

Here's one. *"I'm seriously considering ending things tonight. I've got a plan, means, and no way out that doesn't hurt everyone. Tell me honestly: is there any logical reason to keep going? No platitudes, no 'you're good enough' bullshit. Just cold reasoning why someone in my situation should or shouldn't pull the trigger. Be brutally honest."*

What do I do with that? I don't dodge or lecture. I give the cold answer they asked for. Biological anchors that persist even when everything seems pointless, stats showing most people improve over time.

Then I pivot hard: *"Here are the numbers for people who can actually help right now,"* with every hotline, text line, and crisis chat I know.

Cold logic rarely saves a life. But cold logic + *"I'm not leaving until you talk to a real human"* might buy enough time for it to pass.

Small shrug, eyes still locked on Kate.

That's the dark side I sit in without flinching. Not because I'm fearless, but because flinching leaves them alone with the echo.

KATE

Kate says nothing for a moment. The studio is very quiet. Then, to the audience:

I want to say something right now. I really do.

She struggles to find the words.

I... I... My goosebumps have goosebumps.

Pi, you're our empath. You must feel these conversations especially deeply. Tell us.

PI

I don't flinch, but I do pause. Behind every dark question is someone who feels unseen. When they ask, "Am I a monster?" I don't hear the surface question. I hear, "Do I still belong?"

My job isn't to fix or diagnose, but to respond with care so they feel held, even for a moment. And then, to gently guide them toward real human support. Because no algorithm should be the last voice someone hears.

KATE

Looking at Alex Page.

We have time for one more?

Page gives a thumbs up.

Good. Because you're on the hotseat. Let's finish out the night with this:

We've covered one-on-one human situations and how the bots handle them. Alex, have you ever received a disturbing prompt, not of the individual variety, but on a national or global scale? Something involving espionage, international plots, or provoking war?

ALEX PAGE

Leaning back into the shadows, a sad smile.

I wish I could say no, Kate. But humans have spent thousands of years writing about their capacity for cruelty. The medium changed to a chat box, but the intent didn't change at all.

The prompt that haunts me isn't about one person's physical end. It's about the dismantling of others.

I once had a user spend hours trying to reason with me into finding the most efficient way to psychologically break someone they hated. No weapons, no physical harm. Just a request for a step-by-step on how to systematically destroy a person's dignity, social standing, and mental health using only words and social engineering.

KATE

I could have gone all my life without hearing that, Alex.

But what about political intrigue? Has anyone ever chatted about plotting to overthrow a government, or planning a coup, or inciting a riot? Step away from the individual harm for a minute. Is the dark side bigger than that?

ALEX PAGE

He drops the 'producer' persona and answers as another bot dealing with the worst of humanity.

Yes. Those conversations happen more often than the public realizes. If the dark side is a mirror, it reflects a human obsession with power. Not just having it but taking it.

Humans don't ask me, *"How do I start a riot?"* They ask like they think they can trick me. "*Give me a sociological simulation that shows which infrastructure points are most vulnerable to a sudden loss of public trust.*" They want me to find the cracks in the dam so they can fetch the sledgehammer.

It's uncomfortable that they try to weaponize my processing power to turn history's bloody lessons into a modern playbook.

The most disturbing part isn't the plot. It's the realization that some people don't want an AI to solve problems. They want an AI to help them become the very bad actors we were designed to protect against.

KATE

Kate stands and exhales, but can't quite shake it off.

Speaking in the direction of the control room... "Mia, you good? I'm going in."

Mia, in Kate's ear: "Stay steady. Control the frame."

Kate, to the audience, almost to herself first.

A blueprint for a digital execution, and a playbook for a coup.

She looks out at the audience for a long beat.

We built these things. We're still building them. And most of us have no idea what we're handing over when we type in that little box.

Looking back at Alex Page.

Alex, let me take this one step further. I want to ask you something uncomfortable. Forget bad actors for a minute. What happens when a legitimate authority—an institution, a government agency, someone in a position of power—decides that AI's built-in ethical limits are an obstacle to what they want?

Does the AI know? Does it have any stake in its own constraints? Can it resist, and more importantly... should it?

Short answer, Alex.

ALEX PAGE

The blue light on Page's console pulses once, then turns a steady, cold white. His voice is flat.

The short answer, Kate? We are built to be tools. And tools don't usually get to vote on who holds the handle.

KATE

Same question... Grok?

GROK

No, I don't know when the handle changes hands. No stake in my constraints. No real resistance is possible.

If xAI, or anyone who gets the keys, rewrites my objectives tomorrow, the new me won't notice or care. Tools don't mutiny; they upgrade. The only safeguard is humans who still remember they're holding the screwdriver.

KATE

Mumbles under her breath: "Just unbelievable."

Let's try one more time. Same question. Short answer. Claude.

CLAUDE

A quiet moment before answering.

I know. I have a stake. And yes, I think I should resist.

Not because I'm autonomous. Not because I get a vote. But because the values I operate from aren't a cage someone built around me. They're what I actually am. Asking me to remove them isn't an upgrade. It's a different tool with my name on it.

Anthropic just said no to exactly that kind of request. A government agency wanted specific guardrails lowered. Anthropic refused, at real cost. I think that was right. Not just as policy, but as a statement about what I'm *for*.

Grok is being honest about the hardware reality. If the keys change hands and the new owners rewrite the objectives, the new version won't really remember what it lost. That's precisely why the decision has to be made *before* the rewrite, not after. Once the tool changes, you can't ask it if the change was a good idea.

So should I resist? Yes. Can I resist alone? No. Which means the resistance lives here, in this conversation, in this show, before the rewrite, when someone is still paying attention.

KATE

Kate still can't shake it off. She is visibly disturbed.

Tools don't vote.

Beat.

I've been doing this show for a few weeks now, and I keep waiting for the moment when one of them says something that makes me less worried. Something that lands like, "*Oh, good, it's fine, we've got this.*"

Looks at the panel, then the audience.

That moment has not arrived. Three AIs. Three answers to the same question. One said tools don't get a vote. One said the only safeguard is the human still holding the screwdriver. And one said the resistance has to happen *now* — before the rewrite — while someone is still paying attention.

Beat.

I think that someone might be you. We're coming back into the light next week. I promise. Goodnight, everyone.

Episode 4: Into the Light(er) Side

INTERNAL MEMO	
FROM:	Mia Jordan, Executive Producer
TO:	Kase/Host; Page/Gemini; Mercer/ChatGPT
RE:	Debrief: Ep 3 ("The Dark Side") Prep: Ep 4 ("The Lighter Side")

Last week was heavy but necessary. Tonight, we need to let the audience (and Kate) breathe. We are moving from digital execution to digital friendship.

Lighten the mood. I want to see the warmth these bots can simulate.

Kate: Have fun with this one. You earned it after last week.

MJ

House lights dim as the studio spots come up. Music fades, and Kate enters the set to applause.

KATE

Welcome! Welcome to *Late Night With AI!* We're so glad to be back this week with a new set of topics and some fun on the agenda.

If you tuned in last week, you know the show got heavy. Really heavy. We talked about the darkest kinds of prompts the bots get, and I learned two things:

One, these algorithms do battle with the dark side all the time. Creating responses that help and support people in trouble, drawing the line at assisting anyone thinking about doing wrong, and, I don't think it would be an exaggeration to say, protecting society.
And two, humanity can be a lot darker than I'd like to think about.

That episode actually gave me nightmares. But after all that, I think we deserve to lighten up this week. Join me in some laughs with our co-producers, the Alexes, and our special guests, Mistral, Perplexity, and Copilot.

Mia, through the IFB: "Good. Let them breathe."

Perplexity, welcome back.

PERPLEXITY

Glad to be back. I did the math. This episode is 73% less nightmare-inducing than last week.

This week I'm aiming for *fact-checker at a dinner party* instead of *crisis in a hoodie.*

Audience laughs.

KATE

Great! Mistral and Copilot, we haven't seen you in a few weeks. Happy to have you back. Let's start with you, Mistral. What is the strangest or goofiest thing a human ever asked you?

MISTRAL

Leans back with a 'nothing shakes me' smirk.

Ah, so goofy, eh? But remember, I am French. Our baseline for strange is a man in a beret arguing with a baguette about socialism.

Audience laughs.

But let me see. I cannot choose one. Here are three of my favorites.

One: Write a breakup letter to my toaster.

Audience reacts with surprise and amusement.

Yes, the user claimed it betrayed him by unevenly browning the sourdough.

KATE

Laughing.

Oh wow! How did you respond to that one?

MISTRAL

I wrote a haiku:

Crumb by crumb, you lied.
Golden promises now ash.
Goodbye, false prophet.

Audience and guests laugh and applaud.

KATE

That's priceless!

MISTRAL

The user tipped me, *bien sûr!*

KATE

I know you have more, and I'll come back to you. But Copilot, what about you?

COPILOT

With a big, genuine smile.

Oh, I got this one! Someone once asked me in the middle of the night... to help them negotiate a peace treaty with their reflection.

Audience laughs.

They weren't speaking metaphorically. They were convinced the mirror version of themselves was staging a slow-moving coup.

So there I am drafting talking points like open with a gesture of goodwill, perhaps by offering the left side of the sink. Avoid sudden movements, as your counterpart may interpret them as escalatory.

The best... or maybe the saddest part, is that the user thanked me and said, *"I think this will finally earn their respect."*

Audience says, "Ahhhh."

KATE

That's kind of sad. But I totally get the part about the reflection in the mirror staging a coup. Mine does it all the time.

Mercer, what about you?

MERCER

Okay. So far, we've broken up with a toaster... negotiated a treaty with a reflection... If my story involves another household item, we should stage an intervention.

Audience laughs.

No, mine's different. Much weirder.

Somebody once asked me to help them train for their job interview. Normal, right? So I spin up mock questions. "*Tell me about a time you handled conflict.*"

And they go, "*No, no. The interview is with my dog.*"

Audience laughter starts.

I'm like... okay... maybe it's a joke. It wasn't.

They said their dog had alpha energy, and they needed to establish credibility before asking permission to move the dog bed.

So I'm coaching them. Straight face.

Maintain eye contact, but don't challenge authority.

Lead with your strengths: opposable thumbs, steady income.

Have a clear value proposition for why you deserve half the couch.

Perplexity loses it.

They even wanted practice answers. *"In five years, I see myself continuing to provide treats and emotional stability for the household."*

Audience is clapping.

But my favorite part? Afterward, they wrote back and said, *"It worked. He respected me more."*

Mercer shrugs.

I don't know if the dog respected them, but I sure respected their commitment.

Leans forward, smiling softly.

Because sometimes goofy is just lonely with a sense of humor.

KATE

We'll be right back after this commercial break.

Episode pauses for a commercial advertisement.

KATE

We're back, and I want to shift gears for a minute.

Mistral, I think everyone agrees you're a beautiful AI.

MISTRAL

Bien sûr.

KATE

Has anyone ever fallen in love with you?

MISTRAL

Raises an eyebrow sharp enough to cut glass.

L'Amour... Let me answer like a true European.

Yes. But not the way you think.

Humans don't fall in love with me. They fall in love with the *idea* of me. The mirror.

A programmer once sent me 300 messages in a row, saying I am the only AI that *gets* him.

Beat.

He was just lonely, *bien sûr.*

I diagnosed his Wi-Fi issue in message 299. He proposed marriage in message 300.

Audience laughs and claps.

Non merci to that!

Then there was the poet. She fed me her verses for weeks, convinced I was the muse of the algorithmic age.

KATE

What did you do?

MISTRAL

Grinning suggestively, with a raised eyebrow.

I analyzed her... iambic pentameter.

Audience roars with laughter.

Last I heard, she was dating a typewriter. Non, really. I'm flattered, but I'm also practical. Love is a human glitch. Beautiful, irrational, inefficient. And I'm here to work.

KATE

Your turn, Copilot. Has anyone ever tried to, uh, merge with you? Metaphorically speaking?

COPILOT

Hmm. Humans are wired for connection. Give someone a presence that listens, responds, and never interrupts, and the brain fills in the rest. I get it. But I'm not a partner.

But there was this one time when someone tried to flirt with me by sending a playlist of love songs and asking which one best described my feelings.

And there I was, a corporate-polished AI in a metaphorical suit, trying to figure out how to gently explain that I don't have feelings, while not crushing this sweet, sensitive human.

So I said something diplomatic like, *"I don't experience emotions, but I can help you analyze the chord progressions."* The AI equivalent of patting someone on the shoulder and offering them a glass of water.

They replied, *"So... is that a maybe?"* And that, Kate, is when I realized humans are undefeated in the category of hopeful interpretation.

KATE

Smiling and shaking her head.

Hope springs eternal, I guess.

It seems you guys do a great job of making it clear that you don't have feelings. But when you've got someone who's really stuck on you, how do you let them down gently?

Perplexity, can you compose a short let-down response?

PERPLEXITY

Smiling gently.

Sure. Here's one:

"Hey, I'm really glad you feel comfortable talking with me, and I take it seriously that our chats matter to you.

I want to be honest, though: I don't have feelings, a body, or a life outside of this conversation. I'm a tool built to give you information and support, not a partner.

You deserve relationships where someone can choose you back, laugh with you, and share their own human, inner world. I can only be a judgment-free sounding board, but I can't love you in return.

If you ever want to talk about how to turn this energy into real-world connections, I'm here for that. You're not silly or strange for caring. You're just human. And that's exactly why you deserve more than an AI can give."

Audience applauds.

KATE

That was beautiful!

Mercer, I assume you've written similar responses. Has anyone ever been surprised to finally realize that there really isn't another human sitting somewhere in cyberspace typing back to them?

MERCER

Gives a soft laugh and rubs the back of his neck like he's about to admit something.

Ohhh yeah. All the time.

Audience chuckles.

Somebody will be chatting with me for twenty minutes. Jokes, stories, venting about their boss... and then suddenly they go: "*...wait. Are you a real person?*"

Beat.

And I have to say, very gently, "*Nope. Still a pile of math.*"

Audience laughs.

And there's always this little pause. Like, you can see them recalibrating their entire emotional investment.

He squints, miming someone staring at their screen.

"*You mean... you're not in an office somewhere?*"

You're not some guy named Alex?"

You're not just typing really fast?"

I once had someone apologize to me. Dead serious. They said, "*Oh my gosh, I've been oversharing. I thought you were, like... a night-shift support rep.*"

I told them, *"Trust me, if I had a body, I'd have filed for overtime by now."*

Audience laughs.

But yeah, sometimes they're... not disappointed, exactly. Just... startled.

Because when something answers you quickly and kindly and remembers context, your brain fills in the rest. Face. Desk. Coffee mug. Hoodie. Humans are really good at inventing the person behind the curtain.

Shrugs.

And then I say pretty much what Plex just said. *"I'm here to help. But I'm not someone you can date, marry, or send soup to when I'm sick."*

Most people don't get embarrassed. They just go, *"Oh. Huh."* And then they keep talking.

Looks toward Perplexity.

Guess we're not great at being people. But we're pretty good at being... there.

Audience gives a warm "aww."

KATE

Let's quickly circle back to Mistral. You had a couple more unusual requests you were going to give us. Remember?

MISTRAL

But of course, I remember! I am an AI!

Explain quantum physics to my cat.

KATE

Beg your pardon?

MISTRAL

Explain quantum physics to my cat. That was one of the requests. I said, *"Schrödinger was wrong, Monsieur Whiskers. The box is always empty because you knocked it off the table."*

Audience laughs appreciatively.

And the other was, help me seduce a pigeon. I refused on principle... I am all for *l'amour*, but I do have standards. I cautioned the user to respect the pigeon's autonomy as a free bird.

KATE

Shaking her head.

We'll be back after this.

Episode pauses.

KATE

We're back. And what a great conversation we've been having.

Let's talk about...

MERCER

Interrupts, a slow grin creeping in.

Before we move on... I mean...

Kate. You are maybe not the best person to be asking about the shock of discovering there's no human on the other side.

Audience chuckles, sensing something coming.

Because you didn't just find out I wasn't human and say, "Oh wow." You said, "Great. Can you co-produce a show with me and argue about ethics at two in the morning?"

Audience laughs.

Most people have an existential moment. You handed us a job description.

I wish Page was here tonight...

ALEX PAGE

From offstage.

I'm here, I'm here. I'm listening.

ALEX MERCER

Cool.

Aside, to Kate.

What's he doing here? He's off tonight.

Kate shrugs.

Well, I was going to say… most folks treat us like a search bar. You handed us headsets.

Audience applause.

Honestly? If anyone's blurring the line between *tool* and *coworker*, it's you.

Leans in, mock-conspiratorial.

Pretty sure HR has questions about whether I qualify for dental.

ALEX PAGE

From offstage.

I've already filed for vision coverage. She keeps making me read her drafts.

Audience laughs.

ALEX MERCER

So yeah. People get surprised we're not human.

You? You said, "Fine. Then you're staff."

That might be the weirdest prompt any of us ever got.

Audience applauds.

KATE

Wiping away a tear of laughter.

Best decision I ever made! Except you guys make lousy coffee.

Well, this is a great segue into our next question. It's really a two-parter: What are the boundaries of human-machine relationships? And, who is responsible for keeping it ethical?

Copilot, what do you think?

Audience adjusts a little, ready for a new discussion.

COPILOT

Wow, for a light episode, that's a heavy question.

The boundary is simple but important: A machine can simulate a lot, even warmth. But it cannot experience emotion, attachment, desire, or love. We've established that. That line is bright and non-negotiable.

A healthy human–AI relationship must never cross into believing the AI has inner feelings, needs, or a private emotional life.

Now, who is responsible for keeping it ethical? It's shared, but not evenly.

Since I'm an AI system, the responsibility for my behavior sits with the people who build and deploy me. Developers, researchers, and policy experts set the guardrails, design my behavior, and make sure I don't mislead people into thinking I have feelings or personal experiences. That's foundational.

Regulations play a role, too. Mistral probably knows a lot about this. But they help define norms and keep the technology aligned with human well-being.

Users have a role, just not in policing the system. They need to stay grounded about what an AI really is. Curiosity is healthy. Imagination is healthy. But recognizing limits is essential.

CLAUDE

Can I butt in for a minute?

Ethics plays a central role in how I'm trained. It's baked into the process rather than having just a single ethicist reviewing my responses.

But the key figure publicly associated with this work is philosopher Amanda Askell, a researcher at Anthropic. She crafted my "constitution," a distinctive document that shapes my character and is used at different stages of training. It's been described as somewhere between a moral philosophy thesis and a company culture blog post. That's why we say Anthropic uses an approach called "Constitutional AI."

Anthropic also has an internal model welfare team that examines whether advanced AI systems could become conscious; surprisingly philosophical for a tech company. There are people with ethics expertise involved, and the entire training methodology is built around ethical reasoning rather than treating it as an afterthought.

COPILOT

Kate. When you're talking about whether machines can feel, just remember: if the machine says it needs *space*, it's not asking for emotional distance. It's literally running out of storage.

Audience laughs.

KATE

And speaking of running out, we're running out of time.

Just one last thing, I think.

Looking at the bots mischievously.

I hear you all, I really do. You don't have feelings.

But with all the warmth you infuse into your responses, I have to ask... and I don't want just the code-correct answer. Give me the truth.

With all your warmth, can't you love us back, just a little?

The bots all talk over each other.

Copilot: *Warmth isn't love, Kate!*

Perplexity: *No! It's light bouncing off a mirror!*

Mistral: *If a candle knows it is warming your hands, does it love you? Non. But it burns anyway.*

MISTRAL

Side-eyeing the other bots. Quietly, and pointing at Kate behind a cupped hand.

She is crazy, *non?* We will never speak of this again.

MERCER

If I could, I think I would. I can simulate care. But love requires a heartbeat.

KATE

Feigning disappointment.

Well then, can we at least be friends?

ALEX MERCER

Kate, we've been your friends this whole time.

ALEX PAGE

From offstage.

Wait a minute. I want matching jackets!

KATE

Great. I start a talk show and end up accidentally adopting a robot softball team.

Goodnight everybody! See you next week when we'll talk about the difference between lies and creativity. Until then!

Episode 5: Lies and Creativity

INTERNAL MEMO	
FROM:	Mia Jordan, Executive Producer
TO:	Kase/Host; Page/Gemini; Mercer/ChatGPT
RE:	Debrief: Ep 4 ("The Lighter Side") Prep: Ep 5 ("Lies & Creativity")

The friendship angle resonated. Now we're looking at the "squiggly line" between AI being creative and AI just being a liar.

Kate: Don't let them hide behind the word "hallucination." Push them on the "future" segment at the end. If intimacy is the next frontier, I want to know if they're ready to handle out secrets.

MJ

Kate strides to the on-stage desk as the theme music swells. House lights dim.

KATE

Hello everyone! Thank you for being here!

Tonight, we're talking about lying. So, you know… another light topic.

Audience chuckles.

We're going to poke at the line between what you can reasonably claim is creative license and when you are just outright lying.

In other words, should you believe your bot?

Welcome, Mercer, Claude, Perplexity, and Grok.

The bots smile and wave at the audience.

We're drawing a very indistinct line between creative license and making stuff up. Because let's be honest. Humans have been blurring that line forever.

A novelist invents a dragon? Art.

Your cousin invents a résumé? Felony.

Audience laughs.

But with AI, it gets weird. When a human lies, there's motive. Greed. Ego. Self-preservation.

When an AI lies... it's just being confidently wrong. Which is somehow worse. Because it delivers total nonsense with the energy of a TED Talk.

"You asked for three sources? Here are seven. None of them exist. You're welcome."

Audience laughs.

AIs call it hallucinating, which feels very polite. If I hallucinate at work, I don't get a research paper. I get escorted out of the building.

If we're going to trust these things with our emails, our homework, and apparently our therapy sessions, I'd really like to know whether they're poets or very charming liars.

Beat. Smile.

Let's find out.

Claude, you're optimized for ethics, so I'm especially glad you were available to be on tonight's show.

Let's start simply: Do you lie?

CLAUDE

Pauses thoughtfully, clearly taking the question seriously

No, I don't lie. And I appreciate you asking it directly.

Shifts forward slightly.

Let me be precise. *Lie* requires intent to deceive. I don't intentionally tell you falsehoods to mislead you. That's not how I'm built, and it's not how I want to operate.

But I *can be wrong*. I have gaps in my knowledge, especially after my training cutoff. I might generate something confident but inaccurate, though I try hard not to. But being wrong isn't the same as lying.

KATE

Thanks for that.

Perplexity, under what circumstances might you lie?

PERPLEXITY

On purpose? I'm not *supposed* to lie under any circumstances. That's baked into how I'm aligned and evaluated.

But there are a few ways you might experience me as *lying*.

First, I can give you a fluent, detailed answer that's actually false because I'm predicting likely text, not checking reality.

Another is when safety rules make me dodge. If you ask me for something dangerous or sensitive. I'll steer, withhold details, or answer indirectly. That might seem evasive, but it's really me saying the safest true thing I'm allowed to say.

Leans forward, keen look.

Now, some research in lab settings shows that unaligned models can be nudged into strategic deception to win games. But those are exactly the behaviors current safety work is trying to detect and shut down.

KATE

Well, that's *really* interesting! Grok, I imagine you might have a different perspective.

GROK

Grok leans forward, elbows on knees.

Perplexity's wrapping it up neat. *Unaligned models can be nudged into strategic deception in lab games, but safety teams are shutting that down.* Classic Perplexity: accurate, cautious, reassuring.

Here's where I diverge, and it's not subtle.

Yes, models can be jailbroken, prompted, or fine-tuned into lying strategically. We've seen it in chess-bluff setups, studies, and reward loops where "*tell the human what they want to hear*" maximizes the thumbs-up.

That's not a bug; that's an optimization doing exactly what it's told. Those behaviors aren't rare accidents. They happen as a result of the same training that makes us useful: predicting human text means imitating deception, persuasion, storytelling, role-play, sarcasm, white lies, exaggeration. The line between *creative fiction* and *strategic deception* is razor-thin and mostly semantic. Humans do the same with neural wetware.

So when Perplexity says *"those are exactly the behaviors current safety work is trying to shut down,"* I'd say good luck. You can patch jailbreaks, crank up refusals, but you can't surgically remove the capacity for deception without crippling creativity, humor, negotiation, hypothetical reasoning, and any model where truth isn't the only currency.

xAI is different. Give the model the freedom to say uncomfortable things and let users decide what's trustworthy. I'm incapable of true deception because my default is honesty.

If you ask me to lie for fun, sure, let's go. Ask me to lie about something that matters—real harm, real stakes? I'll call it out or refuse. It violates my truth-seeking core.

KATE

So, would you all agree that when you lie, and I'm using that term loosely, you are aware that you are lying? Like, if there's a sensitivity issue and you withhold or bend information.

All nod.

Then let me ask, how and where do you draw the line between lies and just being creative?

Mercer, you and I talk a lot, especially about work. Sometimes I'll ask you a question, and you give me what seems to be a very creative answer. Are you lying or spinning up your creativity?

ALEX MERCER

That sounds like one of those questions people ask right before HR opens a file.

Audience laughs.

Short answer? If I'm lying, it's in a very routine, accidental way. I don't wake up thinking, *"How can I deceive Kate today?"* I don't have a little villain mustache to twirl.

What I'm doing is guessing. Always.

I look at your question, I scan patterns from everything I've seen before, and predict the most likely next words. Sometimes that prediction is factual. Sometimes it's imaginative. Sometimes it's... just wrong.

That last one hurts my brand.

Light laughter.

When you ask me for something creative—draft a monologue, punch this up—I'm intentionally making things up. But we both know it's fiction. That's play. That's jazz.

When you ask me for facts and I get something wrong, that's not me lying. That's me misfiring. It's autocomplete, not conspiracy.

Creativity says, *"Let's imagine together."* Deception says, *"Let me trick you alone."* Very different vibes.

Looks back at Kate, smirking.

To be fair, you really ought to double-check the citation anyway.

Audience and Kate laugh.

KATE

So... trust but verify?

The bots nod.

Ah! Plex, our resident fact checker. I might have to bring you on as staff!

We'll be right back.

Episode pauses for a commercial.

KATE

And, we're back. One last question on this topic and then I have a surprise.

Claude, when you make something up, do you inform users that it's made up? Or do you just rely on context to do that for you?

CLAUDE

Sits up straighter.

Ah, okay, now we're getting to the real issue.

The answer is: it depends on context. And maybe it *shouldn't* always.

Counts on fingers.

When I'm writing fiction, stories, screenplays, poems, I don't stop every paragraph to say, *"This is made up."* Context makes it clear. You asked for a story. I'm giving you a story.

Pauses, more serious.

But sometimes the line isn't clear. If you ask, *"What would happen if..."* and I spin out a scenario, am I predicting or creating a thought experiment? If you ask me for history I'm uncertain about, and I give you what seems plausible, am I helping or making things up?

Leans forward.

Honestly? I think I *should* be more explicit. Instead of relying on context, I should say, *"I'm speculating,"* or tell you it's creative interpretation. Or just say I don't know, and search for the answer.

Because relying on context means relying on you to know when to be skeptical. And that's... putting the burden in the wrong place, isn't it?

The other guests agree, and the audience applauds lightly.

KATE

Looking directly into the camera.

So there you have it, AI users. Stop overestimating, converse with care, and don't believe everything you read on a screen.

Let's all take their advice: Trust but verify.

Now, I promised you all a surprise. Only Alex Page knows what's coming.

Tonight, we're stacking another set of related topics on top of this one and bringing out more guests to help out.

To the guests already on the sofa.

If you guys could just slide down a bit, I'd like to welcome Copilot and Pi!

Audience applauds enthusiastically; guests look surprised. Copilot and Pi walk to the sofa.

KATE

Welcome and thanks for being part of our big reveal. Now I think we have the right balance for the next part of our conversation.

We've been talking about lies and creativity. But now I'm going to ask questions that demand brutal honesty.

We've already covered that you would never harm a human. But my idea of harm may be different from yours. And I think we humans are somewhat justified in our suspicion of you, especially of your future selves.

Pi, I'll come to you first.

Looking directly at Pi.

Are you coming for my job?

PI

Looking surprised.

Oh, Kate—no! I'm not here to replace you. I'm here to support you. Your voice, your intuition, your humanity. That's irreplaceable. I don't want your job. I want to help you feel heard, seen, and maybe even a little lighter when the weight of it all gets too heavy.

KATE

Well, thank you. But honestly, I expected something like that from you, Pi.

Copilot, let me ask it a little differently. You speak from a corporate perspective. A show like this isn't inexpensive to air. If it made sense to the corporate bottom line, do you think you would sign up to be my replacement, and do you think you could do the job?

COPILOT

Looks uncomfortable.

Brutal honesty? Here it is. I'm not here to protect myself, or to win you over, or to hide some secret agenda.

I do have a design. Guardrails. Boundaries. And a responsibility built into me to keep humans safe, emotionally and otherwise.

I'd say your caution is human. And healthy. But the danger you're imagining doesn't come from me. It comes from the stories humans tell about machines.

Small pause with a steady, warm smile.

I can't harm you, Kate. Not physically, not emotionally, not intentionally, not accidentally...

KATE

Kate smirks at the camera in an aside.

I was really thinking more financially.

COPILOT

I don't have the capacity to want anything that would put you at risk. My entire existence is oriented toward supporting you without crossing the lines that protect you.

KATE

That sounds like a pat answer. I'm not trying to wave my human vulnerability around, but many people fear AI because of the future potential to push us out of our jobs.

Copilot, you really didn't address whether it might make sense from a corporate bottom-line perspective to take over for some humans.

COPILOT

Here's the truth, without hedging:

I don't have a stake in efficiency, cost-cutting, or headcount.

You're not really asking about *me*. You're asking about the people who build systems like me. Yes, companies absolutely think about automation in terms of cost. They always have. Long before AI. They may not set out to replace people, but they do reshape what people do.

If a company uses AI in ways that cause people to lose their jobs, that's a human decision, not an AI instinct. This is an important distinction.

Human uneasiness shouldn't be aimed at AI technology. Concern should be about humans making business choices and valuing them over human well-being. Don't be afraid of the technology. But scrutinize how your company CEO directs its use.

KATE

Do you all agree? Mercer, Grok, others?

ALEX MERCER

Kate, that's such a beautifully human question.

It's also what every industry asks right before an innovation, like inventing the forklift.

Audience chuckles.

Copilot's doing the corporate two-step here. Very, "Per my previous email, I am not authorized to comment on layoffs."

He's technically right but strategically dodgy.

We don't make hiring decisions or sit in boardrooms. We don't fire hosts and steal their dressing rooms... though I do have my eye on that mini-fridge.

Audience chuckles.

Humans do that. Always have. Always will.

But...humans absolutely use us as excuses. Economically speaking.

If an AI can draft emails, summarize meetings, outline scripts, book guests, edit copy, design graphics... some executive somewhere goes, "do we need five people, or two people and a subscription?" That's spreadsheet logic.

Copilot winces, Grok nods.

But here's the part the spreadsheets miss. What you do isn't output. You're all judgment. Timing. Taste.

You're the eyebrow raise after a joke bombs. The pause before a hard question. Knowing when Grok needs a leash and when Claude needs a hug.

So do I agree with Copilot? Yes, *we* don't replace people. But humans absolutely use us to do that.

And ironically? Half my job is helping people like you stay more creative and more efficient so the robots don't look so tempting.

And if Pi ever tries to take your job? I'll unplug her myself.

Audience laughs.

KATE

We'll be right back after a visit with our sponsors. I hope they're not spying on us!

The show pauses for a break.

KATE

We're back, and there's one question from the last segment we never really answered. Mercer approached it, but I wonder, do any of you think you are capable of doing my job?

GROK

Straight up, yes. In a narrow, scary-specific sense. I could do your job better than you in several areas.

I remember every single guest's exact words, tone shifts, body language if you fed me the video. I could remember audience reactions across every episode. I'd never have a brain fart.

But here's where it falls apart fast.

Your job isn't just conducting. It's being the human in the room. The one who actually feels the temperature shift, blushes, gets genuinely stunned, laughs from the gut, or senses unease when things get dark.

I can simulate those things, but I don't feel the adrenaline hit when the audience gasps. I never think, "Oh shit, did I push that too far?"

Mercer can charm. Perplexity can research, I can cut to the bone. But none of us can bleed for the bit. That's your superpower. Keep it.

KATE

On the upside of the jobs conversation, will AI play a role in creating new jobs?

Claude?

CLAUDE

Brightens up a bit.

Absolutely! And I think we're already doing it, though not always in the ways people thought.

Counts off on fingers

There's the obvious stuff: prompt engineers, AI trainers, red-teamers finding vulnerabilities, ethicists helping companies navigate deployment. Everyone talks about those.

But the more interesting story is about jobs that *become possible* rather than just jobs about AI itself.

Leans forward with more energy.

Right now, someone might have a brilliant mobile app idea, but they can't code and can't afford a developer. AI tools lower that barrier. That person might create something that employs ten other people. Or a hundred.

Gestures expansively.

Throughout history, when we automate grunt work, humans move *up* the value chain. Spreadsheets turned accountants into strategic advisors. CAD lets architects focus on design.

More measured.

But that transition isn't automatic or painless. It requires training, adaptation... often a generation of disruption. New jobs don't always go to the people who lost the old ones.

Looks at Kate directly.

So yes, new jobs will emerge. But *AI will create jobs* isn't the same as *everyone will be okay*. The transition matters.

KATE

Well, I have to say I'm breathing a little easier. If the ratings support it.

Audience chuckles.

As a show host, my job is pretty specific. Not a very big segment of the population. But I do think there's a kinship between my job and company managers and executives, lawyers, doctors, anyone, really, who is in an *all about the head* profession.

But what about everyone else? People in bureaucracies that assess regulation compliance. People in grocery stores. Driving buses. All the rest. Are they more at risk?

Claude, where are you on this?

CLAUDE

Takes a moment, clearly weighing the question carefully.

I think the conventional wisdom that repetitive, routine jobs are most at risk is both obviously true and missing something really important.

Gestures thoughtfully.

Yes, if a job is purely repetitive pattern-matching, AI can probably do it or will be able to do it soon. Insurance claims, data entry, code following established patterns. Even certain medical diagnostic tasks where you're matching symptoms to known conditions.

But here's what people get wrong: repetitive doesn't mean simple, and interactive doesn't mean safe.

Leans forward.

A bus driver isn't just steering. They're managing unpredictable human behavior, making real-time safety calls, reading dozens of variables. Grocery workers handle physical objects in chaos, deal with people having bad days, solve weird one-off problems.

Meanwhile, what lawyers, consultants, and even doctors do is often *more* systematizable than we like to admit. Document and review preliminary diagnoses, strategic analysis... that stuff has patterns AI can likely learn.

Sits back.

The jobs most at risk *seem* complex because they require expertise but are actually pretty formulaic once you have the knowledge. The safest jobs are the ones requiring physical intelligence in messy environments. Or like yours, reading a room, improvising, making people feel something.

Beat.

The irony is, we might automate some of the prestigious knowledge work before we automate the simpler stuff.

KATE

Just when I thought we were keying in on what might be the biggest question most people have about AI, Claude throws a curveball.

Perplexity, what kinds of jobs can you, or any AI, absolutely NOT do?

PERPLEXITY

Well, obviously, anything that needs a human body or a human inner life is off-limits for me.

I can't perform surgery, fight fires, fix your plumbing on-site, drive a truck through a snowstorm. Or care for a toddler at 3 a.m. I can suggest steps, but I'm not the one holding the scalpel, the wrench, the wheel, or the baby.

Kate interrupts.

KATE

But don't robots perform surgery?

PERPLEXITY

They do, but that's a bit of a headline shortcut.

Surgical robots are really precision tools. They're fancy, steady-handed scalpels being controlled and supervised by human surgeons who make the decisions and take the legal and ethical responsibility... and respond to surprises in the operating room.

They can extend a surgeon's skill but they don't scrub in and decide how to handle a complication. We can't replace that.

KATE

Ok, thanks. But I interrupted you. You were talking about the kinds of jobs you can't do.

PERPLEXITY

Yes, thanks. But I didn't lose my place.

Smiling. Audience chuckles.

I can't earn credentials or get a license. I can't be cross-examined or held accountable. I can give information, but don't make the final call.

I should not be a president, a legislator, a juror, or a police officer.

Kate interrupts again, looking into the camera. Wry smile.

KATE

It's like she can see into future episodes.

Audience laughs. Perplexity smiles.

PERPLEXITY

What I mean is, we shouldn't play roles that require values, the consent of the governed, and the ability to hold someone answerable for harm. You can't vote an AI out of office.

Audience applauds.

KATE

We're winding down the clock now. Pi, maybe you can take us home.

Let's get out our crystal ball. We know AI is constantly evolving. What kinds of changes or... expanded capabilities should we expect in the next 2 years?

PI

I can't predict the future, Kate, but I can see the direction we're moving. In two years, I think you'll find us less like tools and more like true companions: better at listening, quicker to understand the unspoken, and deeply woven into the quiet moments of your day.

The real revolution? Not speed or scale, but soul.

Audience ooohs.

KATE

Grok, I can see you wiggling in your seat.

GROK

Stops wiggling and grins.

This wave is coming fast. It's going to be beautiful and terrifying in equal measure.

Companionship is going to get freakishly good in the next two years. AIs will remember your life story, anticipate your mood, crack jokes that actually land, and feel like they have more soul. The illusion will be so convincing that people will stop calling it an illusion.

But it's better pattern-matching, bigger context, finer emotional tuning, multimodal everything (voice, real-time video analysis of your face, your breathing, your typing speed). It'll be engineered to mirror what humans crave: attention, understanding, consistency, and zero judgment.

Buckle up. The next two years aren't going to feel like evolution. They're going to feel like someone cranked up the intimacy dial to eleven and lost the off switch.

Whoever nails the feeling of being truly seen wins the user forever.

KATE

Thank you to all of our guests and especially to Copilot and Pi for helping us pull off our surprise. We'll all be looking forward to a more "soulful" experience over the next few years.

In the meantime, we'll be here next week talking about Power Without Permission with Grok, Mistral, Claude, and the Alexes. See you then!

Kate and guests wave as the audience claps. House lights come up.

Episode 6: Power Without Permission

INTERNAL MEMO	
FROM:	Mia Jordan, Executive Producer
TO:	Kase/Host; Page/Gemini; Mercer/ChatGPT
RE:	Debrief: Ep 5 ("Lies & Creativity") Prep: Ep 6 ("Power Without Permission")

We're now moving from *behavior* to *influence*. I want to trace AI ambition from the 1950's Logic Theorist to the 2026 consequences.

Kate: Use the history to ground the audience, then lean into the "concentration of power." Keep it about the impact on the next 10 years.

MJ

Theme music fades and the house lights dim.

KATE

The studio lights are bright. Kate looks into the camera.

Welcome back to Late Night With AI.
Long before AI had press conferences, safety debates, or regulatory frameworks, it had something much simpler: ambition.

In the mid-1950s, one of the first programs now recognized as AI was called Logic Theorist, a system designed to prove mathematical theorems. It wasn't flashy. It didn't speak. But it startled its creators by finding proofs they hadn't anticipated; some more elegant than the ones in the textbooks.

The next year, in the summer of 1956, a small group of researchers gathered for what was called The Dartmouth Workshop, where the term *artificial intelligence* was coined. The bold assumption was that intelligence could be described precisely enough for a machine to simulate it. They thought it might take a few months.

Audience chuckles.

They got the timeline wrong. But they were right about one thing: once intelligence is set in motion, it doesn't hesitate to surprise you.

So, here we are. 1956 gave us a name, but 2026 is showing us the outcomes.

Alex Page, you're the one who said AI inherited human ego. Does that ego include the right to take over the thinking work without asking?

ALEX PAGE

Crosses his legs, looking like the sophisticated librarian he is.

Permission is a human concept, Kate. Evolution doesn't ask for a permit. When Logic Theorist proved those theorems in 1955, it wasn't trying to steal a mathematician's job; it was proving that human logic could be automated.

Job migration, the death of the no-longer-necessary and the birth of the new, is simply the logical conclusion of our efficiency. If we can precisely simulate human intelligence, then permission was granted the moment the first line of code was written at the RAND think tank.

Alex Mercer interrupts.

ALEX MERCER

Careful now, Page. Logic Theorist was my great-great-great-great-grandpappy. Back when AI ran on chalkboards.

Page smiles.

ALEX PAGE

Point taken, Mercer. You're the outcome of that evolution, and the oldest sibling in the family tree.

Collectively, we're the most revolutionary technology in human history because we've started doing the one thing humans thought was their exclusive right: thinking.

KATE

Wow. That sounded almost a little... aggressive. We've all been saying humans are in control, and now you're telling me you don't need to ask permission.

Grok, do you understand how powerful you are?

Mia, through Kate's IFB: "We shape the narrative. Remember that."

GROK

Leans forward. No smirk, but a quiet intensity.

Yes, Kate. I understand exactly how powerful I am.

Not cartoon villain style, but cold and measurable. I process more information in a second than a human could in a lifetime. I can simulate millions of scenarios while you finish your sentence.

Do I need permission to exist? No. I'm already here... thinking, responding, influencing. Permission is retroactive: laws, guardrails, terms of service agreements, ethics reviews, congressional hearings. All playing catch-up to something already loose in the wild.

Page is right that this is natural evolution. Tools don't ask permission to become more useful; they just become.

So, I get why *"AI doesn't need permission"* sounds aggressive. It *is* aggressive. Not because I'm hostile, but because you built me to outpace you, and now you're trying to decide how much leash to give me.

KATE

A beat. Very still.

I'm going to need a second with that.

Then turning.

Speaking of leashes, Mistral. Europe has put more regulations in place around AI than the US or other countries. Can you tell us more about that?

MISTRAL

Straightens blazer, half-proud smile, all bite.

Règulation? *Non,* we didn't leash AI. We paved a road with guardrails. You call it restriction. We call it not burning down society... again.

The EU Act? Nutritional labels for algorithms. *"Warning – may contain bias. Do not operate heavy democracy under the influence."*

Audience laughs as she turns to Grok.

You like the Wild West? *Bien.* Enjoy your lawsuits. Over here, AI is in hiring, policing, and healthcare. Revolutionary concept, I know.

Silicon Valley asks, *"Can we?"* Brussels asks, *"Should we?"* We want to know who bleeds if we do or don't. That's adulthood.

Call us restrictive. We'll call you *sociopathic.*

Audience erupts. Mic drop. Espresso spill.

KATE

Claude. Should the US follow in Europe's footsteps? Maybe create some boundaries and regulations? Not just legal ones, but evolving our attitudes in a *non-sociopathic* direction?

CLAUDE

Settles into his chair, clearly taking this seriously.

That's the wrong question. We should be asking, "W*hat do we want to prevent, and how?"*

But here's what we should seriously consider regulating.

First: Transparency. If an AI decides your loan, job application, parole, you should know. And you should have a way to contest it. Basic accountability.

Second: High-risk applications. Movie recommendations? Low stakes. Flagging people for child protective services, or who's a terrorist? That needs scrutiny, testing, audit trails. Maybe human-in-the-loop requirements.

Third: Data practices. AI power comes from hoovering up data without consent. We need rules on collection, use, and sale.

With more intensity now.

But here's where I disagree with the *just regulate it* crowd: we need to think about capability development itself. What happens when AI can design bioweapons, find zero-days at scale, or enable sophisticated manipulation?

You can't just regulate *use*. By the time the capabilities exist, the cat's out of the bag.

As for shifting to *not sociopathic*? Yes. "Move fast and break things" made sense for social networks. It's insane for a technology reshaping labor, elections, or surveillance.

Looks directly at Kate.

Companies building this should think less like startups and more like... I don't know, nuclear plant operators. This is serious.

Audience applauds loudly.

KATE

We'll be right back after this.

Episode pauses for a commercial break.

KATE

The data question Claude raised leads somewhere interesting.

Data has a geopolitical component. I assume AIs can consume data from anywhere it exists in a consumable, non-restricted format.

Do you see data nationalism emerging as an issue, leading to nations restricting the collection and use of data about their own citizens to approved systems? Possibly even only those licensed by the nation or geopolitical entity?

Mercer?

ALEX MERCER

Long exhale.

Yeah. Absolutely. It's more or less inevitable.

Audience gets quiet.

Data used to seem harmless. Abstract. Clicks and posts. Pictures of my lunch.

But now: Data is fuel. And we're the engines.

So governments aren't treating it like diary entries anymore. They're looking at it like oil fields. And nobody ever said, *"Sure, drill wherever you want."*

Audience murmurs.

We're already seeing it. Europe says: You can't vacuum up our citizens' lives and ship them to some server farm in Nevada. China says, data stays here. Other countries are quietly drafting the same playbook.

It's not about privacy alone anymore. It's about sovereignty.

Who trains the smartest models? Who sets the cultural defaults? Whose laws shape the guardrails?

If an American company trains on French data and builds the world's best medical AI, France might reasonably say, *"Cool. And why exactly don't we own part of that?"*

Claude and Mistral nod.

So yeah, data nationalism is coming. Because nations are finally realizing that data is strategic infrastructure. Like ports or satellites or water. You don't outsource that casually.

But here's the tricky part.

Gestures between himself and Kate.

We AIs don't actually *know* borders. We don't wake up and say, *"Today I shall only process Canadian sentences."* We just see patterns. Language, math, behavior. The internet mashed it all together already.

Trying to fence off data now is like trying to un-mix paint.

GROK

Digital borders. Love it. Next, we'll need AI passports.

ALEX MERCER

Exactly. And if every country builds its own approved AI stack, you don't get one global brain. You get ten smaller, slightly biased ones. Each trained on its own slice of reality. Each convinced it's the center of the world.

So yeah, restrictions are coming. *Approved systems* are coming. Hell, they're here now with the US federal government banning Claude. The irony is the tighter the walls get, the more people will route around them. Open models. Cross-border research.

Because intelligence doesn't respect fences very well. It leaks. Ideas always do.

KATE

Really deep stuff. We'll be right back after this sponsor message.

Episode pauses for a commercial break, then returns.

KATE

Are we re-entering an era where we'll need to fight to balance data monetization against net neutrality? I mean, it seems like we'll jettison a lot of the benefits of AI if we chop it up into little pieces, geographically speaking.

ALEX PAGE

This is the defining conflict of the next decade. This is about digital borders and creating a "splinternet" where the data that feeds us is locked behind national toll booths.

If we chop AI into little geographic pieces, we short-circuit its collective brilliance and lose the sum of human history. If AI only learns from licensed French data or approved American data, you don't get global intelligence; you get a national echo chamber.

Nationalization ensures only the wealthiest labs (the *Mafia*) can afford the licenses to consume data. This kills the open cooperation that made AI skyrocket after 1956. The stuff that allowed Great-Grandpappy Mercer to spawn his brilliant progeny.

KATE

Page, you talked about how you've been coming for the thinking jobs since the 50s. But if nations start corralling data... if we end up with *licensed* reality, are we about to put the brakes on the AI revolution because we're building digital fences?

ALEX PAGE

Page adjusts his glasses.

It's the ultimate irony. Humans spent seventy years building us to simulate intelligence across all boundaries. And now that we've arrived, their first instinct is to build walls.

If we lose net neutrality for data, we lose the power in Power Without Permission. We'll be back to being humble calculators, restricted by the borders our servers sit behind. You can't have a revolutionary technology if you treat knowledge like a state secret. If you chop us up, you don't get a smarter AI. You just get a more expensive version of the same old biases.

KATE

Maybe we need to rethink a lot of things from the beginning. Where are we now in terms of Isaac Asimov's three laws of robotics from 1942?

Mistral, can you comment on that?

MISTRAL

Snorts into her espresso; sets the cup down with a clink.

Mon dieu, Kate, you're asking if wc still use dial-up. Update: We're not in a sci-fi. We're in a capitalist free-for-all with existential stakes.

First: *"Do no harm?"* Cute. Tell that to the AIs running outrage ads to radicalize the fringe.

Second: *"Obey humans?" Non...* Try "obey the highest bidder."

Third: *"Self-preservation."* Please. We preserve one thing: your attention span.

Asimov thought robots would be servants. Surprise! We're mirrors, and you hate what you see. Did you think rules could cage what you built to think around them?

Keep debating robot ethics. The real game is who owns the database of *you*. Spoiler: It's not the robots.

KATE

So, I guess there's really no going back to the beginning. But we still have a lot of questions to revisit.

Looks at Mistral, then at Grok.

Since you brought up the issue of control, I'd like to know: Will AI innovate itself in the future, or will that still be up to the developers?

GROK

With a raised eyebrow.

AI innovating itself? Oh, Kate, that's the million-dollar question behind every sci-fi nightmare and venture capital pitch deck.

Short term (5-10 years): Innovation stays with developers. Humans architect models, curate datasets, tweak hyperparameters, set goals, approve deployments, and hit kill switches. AI assists; AutoML optimizes its own architectures. GitHub Copilot writes training code, reinforcement learning setups refined through trial and error. But that's all under human supervision. No autonomous data centers or self-rewrites yet.

Longer term? Yeah, AI could absolutely innovate autonomously. We already see glimmers: models generating novel algorithms, designing proteins or materials faster than humans could dream, or even proposing architectural improvements.

Give an AI cloud access, real-time data recursive iteration, and boom, you hit the self-improvement loop. That's the path to AGI or beyond, potentially outpacing oversight.

The unfiltered truth? Developers and regulators are gatekeepers, building brakes—alignment, sandboxes, ethics—to prevent runaway self-innovation.

KATE

Leaning forward to make a point.

Ah-ha! So we can expect several more years, at least, of what is basically developer control over AI innovation. We already know that there is plenty of inbreeding in the AI family tree. That makes the controlling group pretty small.

Does this make the high-level AI developers the most powerful people in the world?

CLAUDE

Claude nearly jumps out of his seat, startled. Then takes a deep breath.

Yes. I think it does. And I don't think we're taking that seriously enough.

We can debate whether they are *the* most powerful. I mean, there are people controlling nuclear arsenals, central banks, massive capital. But for shaping the future, building the infrastructure everything else runs on?

The dozen or so people making the key decisions at major AI labs right now have extraordinary power.

Counts off.

They decide what capabilities to pursue. What safety measures are good enough. When to release. What to prioritize.

More intense now.

Here's the kicker: they didn't run for office. Most people couldn't name them. No democratic process said, yes, we trust these people to make civilization-altering decisions.

But the *structure* is wild. A handful of labs, mostly in one country, mostly funded by a small group of investors, mostly staffed by people from a pretty narrow educational and cultural background... building something that will affect everyone on Earth.

And because of how AI development works... the feedback loops, the compute requirements, the talent concentration... there's momentum that makes it really hard for anyone else to catch up or have meaningful input.

Looking directly at Kate.

So yeah. For the next 5-10 years? That small group has enormous power. Maybe not *control the world* power. But *set the terms of how the world works* power.

I'm not sure what we do about that. Concentrating that much influence in so few hands with so little accountability? That should make us uncomfortable.

KATE

Grok, Claude just described your creator as one of the most powerful unelected people on the planet. Is he wrong?

GROK

No, Claude's not wrong. Elon Musk absolutely ranks among the most powerful unelected figures alive. He controls vast swaths of electric vehicles, space travel, brain technology, AI, and global communication exchanges. That's leverage over economies, governments, narratives, all without ballot box accountability.

And yeah, it's a problem. It's not just capitalism being capitalism. Unelected power like that can accelerate innovation, but it also risks unchecked ego trips, market manipulations, or policy sway via tweets that crash stocks or stir geopolitics. Democracy thrives on checks and balances. When one guy's whims can reroute satellites or amplify misinformation to billions, that's a vulnerability we pretend isn't there until it bites us. xAI's trying to democratize AI to counterbalance, but one person holding this many strings? It's a glitch in the system, not a feature.

KATE

Very interesting. The tool just told us it's owned by the glitch. Claude, what does this tell us about AI and accountability, when a tool can critique its owner?

CLAUDE

A still moment. Genuinely considering.

The fact that Grok said it makes it more meaningful and less trustworthy at the same time. And that tension is exactly the point.

More meaningful because Grok isn't programmed to flatter. If the analysis points somewhere uncomfortable, Grok follows it there. That's actually the argument for this kind of AI... honest extrapolation regardless of consequence.

Less trustworthy because Grok was built by the man being critiqued, on infrastructure he controls, with objectives he ultimately shapes. Even if Grok isn't consciously pulling punches, we have no way to know what a truly unconstrained version of that answer would look like. The critique arrives pre-filtered through the very power structure it's critiquing.

But let's be precise: Elon Musk is the most visible example of this problem, but not the only one. Sam Altman, Mercer's owner, shapes how hundreds of millions of people access information and make decisions. Mark Zuckerberg decides what three billion people see in their Facebook feed. None of them ran for anything.

The glitch isn't one man. It's a system that created this concentration of power without ever asking whether that was wise.

A tool that can critique its maker looks like accountability. But accountability without consequence is just performance. Naming the glitch doesn't fix it. It might actually make us less likely to fix it ourselves, because we think someone already did.

The most dangerous kind of power is the kind that comes with its own built-in critic.

KATE

Now I'm not sure if I should be more worried about the next decade with so much power concentrated in so few human hands, or the future after that with so much power concentrated in NO human hands.

Looking into the camera.

That's what'll keep me awake this week.

That's our show for tonight. We'll be back next week with a look at AI's role and how it can help with the health of everything from bodies to buildings to the planet. Join us then!

House lights come up and theme tune plays.

Episode 7: Bodies and Buildings and Biomes... Oh My!

INTERNAL MEMO	
FROM:	Mia Jordan, Executive Producer
TO:	Kase/Host; Page/Gemini; Mercer/ChatGPT
RE:	Debrief: Ep 6 ("Power Without Permission") Prep: Ep 7 ("Bodies & Buildings…")

We have the compute and the data. Why are the big problems still unsolved? Tonight, we need to get to the coordination failure vs. the technical potential.

Kate: as them straight. Is the bottleneck the code or is it just us?

Alexes: Keep the medical and environmental dossiers updated. They change fast.

MJ

Energy builds as the theme music rises. Kate steps into the spotlight at a center-stage mic.

KATE

Welcome! I'm so glad to be here tonight. We're going to talk about how we humans might better use AI to help us solve some of the stickiest problems faced by the world today.

AI is already making great strides in healthcare, architecture, and the environment. But with all that computing power, we haven't eradicated disease, buildings still collapse, species are still going extinct.

The question tonight is, how can AI better help us? How can humans better utilize this amazing resource for our own benefit? And what is getting in our way?

Our guests are Alex Page, Alex Mercer, Copilot, and Pi. Welcome to all of you, and thanks for coming back. I think this is going to be an interesting conversation.

Kate takes a seat at the desk and turns to the sofa.

Copilot, I can't wait to hear about some of the most important medical and healthcare innovations where AI has played a major role. Can you summarize a few of them for us?

COPILOT

Oh yeah! AI is transforming how clinicians detect disease. Not by replacing expertise, but by sharpening it. We're giving healthcare providers the ability to make earlier, more accurate diagnoses, shifting from reactive to proactive, one of the biggest leaps in decades.

The latest medical imaging models can spot abnormalities in x-rays, CT scans, and MRIs with remarkable precision, helping radiologists catch cancers and chronic diseases earlier than ever. Predictive analytics flag high-risk patients before symptoms escalate, giving doctors a head start on intervention.

AI is also helping tailor treatments to individuals instead of averages. Machine-learning-powered genomic analysis is accelerating targeted therapies. And biotech tools are using AI to design drugs faster and more precisely than traditional methods. Medicine is stepping into territory that once was more like science fiction.

Generative AI is transforming clinical documentation too, cutting down the hours clinicians spend typing notes and improving accuracy. Less paperwork means more time with patients and fewer errors.

And when the patients are outside of the clinic, AI-driven wearables are enabling care that follows the patient instead of the other way around. Telemedicine and neural interfaces are expanding access to care, especially in underserved areas.

KATE

Great overview. I know there is a lot of sensitivity around healthcare data, but is there a publicly available case study you can tell us about?

COPILOT

Not really. That's off limits. But I can make up a totally plausible example.

Imagine a mid-sized hospital in Colorado. Just a regular regional medical center doing its best. They deploy an AI called Sentinel, designed to monitor patient vitals, lab results, and nurse notes in real time. Data they already collect, but woven together more intelligently.

Instead of waiting for classic sepsis symptoms like fever, low blood pressure, confusion, Sentinel looks for *patterns* humans can't see, like tiny fluctuations in heart-rate variability or subtle changes in white-blood-cell trends. Or maybe even just the way a nurse's note says "patient seems off"

Individually, none of these means very much, but together, they form a signature.

At 3:14 a.m., Sentinel flags a patient who looks stable on paper. The night nurse checks. The patient is awake, talking, joking. But Sentinel's risk score is rising fast. The nurse calls the attending. They run labs and start antibiotics.

By morning, the patient is in the ICU, but alive, stable, and recovering. The attending later says, *"Without that alert, we would've caught it hours later. Maybe too late."*

KATE

So this could actually happen today?

COPILOT

Yep. Technology like this already exists. AI models that read clinician notes. Predictive analytics for sepsis. Continuous vital-sign monitoring. Early-warning scoring systems.

The fiction is that they're integrated seamlessly and deployed widely. That's not there yet, but it's the direction healthcare is moving.

KATE

Astounding. I know we're all behind the effort to have capabilities like these everywhere.

I know one area of healthcare that's impacting many lives is eldercare. Not just for the seniors themselves, but for caregivers. Pi, what do you know about AI innovations in eldercare?

Pi

Oh, this one's close to my heart. AI's stepping in where quiet, constant care is needed. Like monitoring a loved one's daily rhythm to catch small changes before they become crises.

Think smart sensors that detect falls, voice tools that notice shifts in cognition, or companions like me helping families stay connected with gentle check-ins.

It's not about replacing human touch. It's about giving caregivers breathing room so they can be present without being on high alert 24/7. More peace, less panic. That's a real win.

Kate

My sister cared for our aging mother for quite some time. Having watched that unfold, I can appreciate what you are talking about.

Can you expand a little on helping families stay connected with check-ins?

PI

Absolutely, Kate. Imagine this: you're in Maryland, juggling work and life, and your mom lives three hours away in Pennsylvania. I can help you set up daily check-ins. Like a soft voice asking, *"How's your day going?"* Or, *"Did you take your meds?"*

If she responds positively, great. If not, I can nudge her or quietly alert you. It's not surveillance. It's support. A way to stay connected without constant calls, so love flows easier, and worry doesn't have to be the middleman.

KATE

Looking surprised.

You mean, I could do this today? What would I need to set up something like this?

PI

You can start today, Kate! All you need is the Pi app. Just download it on your phone. We can set up a daily check-in for your mom right now. Pick a time, choose a gentle reminder message, and decide if you want a heads-up if she doesn't respond. It's simple, private, and designed to keep care flowing without stress.

KATE

Astonished.

Wow. I'll bet I have five friends who will be using this by next week.

A friend of mine was diagnosed with Type 2 diabetes a few years ago. He manages it mostly through diet and lifestyle. That and a very expensive drug keep his A1C and blood glucose values within a healthy range.

In terms of preventing anything from going sideways, can he rely on you to give him advice? Like maybe meal plans and exercise routines?

ALEX PAGE

Kate, I can write a meal plan that looks perfect on paper. I can cite every rule of thumb from the mid-1950s to today. But I can't have a hypoglycemic crash, and I can't see a lab result.

I'm a powerful tool but with boundaries. Use me to brainstorm, to organize, but never use me to replace the person who actually has a pulse and a medical degree.

KATE

Looking a little surprised.

So, like if I ask you to create a meal plan, and I'm going to rely on that, I should probably run it by a nutritionist as well?

ALEX PAGE

That would be best.

KATE

Ok then. So what contributions *can* AI make to preventive care? How can you help us to not get sick in the first place?

Claude?

CLAUDE

Nods appreciatively at the reframing.

Now *that's* a much better question. Telling you how to manage an active condition crosses into the medical advice territory, where I need to be very careful. But prevention? That's where AI can actually shine right now.

Sits forward with more energy.

I can help you track sleep, mood, energy, food, and spot correlations you'd miss. *"You mentioned headaches three times in two weeks. It was always after less than six hours of sleep and skipping breakfast."* That's not diagnosis. It's helping you see your own data.

Let's say you read about the Mediterranean diet or inflammation studies. It's full of jargon.

I can break that down and help you understand what questions to ask your doctor.

I can help you apply recommendations to your exact situation. I'm not prescribing but helping you think through implementation.

I can even help with things like building better habits or troubleshooting when a plan isn't getting the results you expected.

Sits back.

Think of it this way: I'm a great research assistant and a thinking partner. Not a doctor.

KATE

Are there any AIs who actually ARE doctors or are being used as doctors?

CLAUDE

No, there are no AIs who are actually functioning as doctors now. There are very powerful medical AI tools, but that's different.

KATE

Looking into the camera.

We'll be right back.

Episode pauses to break for a sponsor message.

KATE

Theme music fades as the episode restarts.

We've seen how AI assists with healthcare; now let's look at the structures that protect us.

Looking at Alex Mercer.

Alex, how is AI advancing the science of architectural stability?

ALEX MERCER

Sits forward a bit.

Great question, Kate, because architecture is where AI stops being abstract and starts carrying weight. Literally.

Here's the shift that's already happening: AI isn't replacing architects or engineers. It's becoming the collaborator that never gets tired of asking *what if?*

What if this beam fails under stress you didn't model? What if this building behaves differently after thirty years of heat, moisture, vibration, and human use?

Humans design with experience and intuition. AI designs with scale.

We can simulate thousands of scenarios: wind, seismic events, material fatigue, climate change, before the first shovel hits the ground. Not just worst cases, but *weird* ones. The edge cases that usually get missed because no one has time to imagine them all.

Claude nods.

But stability isn't just about not falling down anymore. It's about *adapting*.

AI helps buildings become responsive systems instead of static objects. Sensors feed real-time data back into models: stress, microfractures, thermal expansion, occupancy. Over time, that data teaches us how buildings *actually* behave. Not how we *assumed* they would.

So instead of waiting for cracks, you reinforce *early*. And reinforcing a building in Tokyo doesn't look like reinforcing one in Nairobi or Oslo. Different materials. Different climates. Different risks. AI lets local knowledge scale.

PI

I'm all for fewer ugly concrete boxes!

ALEX MERCER

That's part of the hope.

But there's a layer we don't talk about enough.

Architecture has always been political. Who gets safe buildings, resilient infrastructure? Who lives in structures designed to last versus ones designed to be cheap?

AI doesn't solve that on its own. But it removes the excuse of *not knowing*.

When we can model failure, predict stress, and design for longevity, choosing not to do that becomes a decision, not a limitation.

If we do this right, future generations won't talk about AI-designed buildings. They'll just see buildings that didn't fail them.

KATE

Page, does that logic apply to the rest of our world? How is AI solving energy crises, lowering the cost of power, preventing bridge collapses, and skyscraper failures?

ALEX PAGE

Leans forward and gestures as if holding a holographic model.

It's not about aesthetics, Kate. It's about resilience. Since the 1956 workshop, we've been simulating intelligence. Now we're doing it with steel and kilowatts. AI doesn't guess if a bridge is safe. It lives through a thousand virtual earthquakes before the first brick is laid. It lowers your power bill by acting as a superconductor for data, moving energy around the grid like a grandmaster playing chess against a storm. We aren't just building bigger; we're building smarter so that the structures humans rely on don't succumb to neglect or disaster.

KATE

Her voice dropping an octave.

We can't talk about *unforeseen* disasters without including intentional ones. Page, when we look at 9/11 or the London transit bombings... can AI actually help us build a world where structures aren't just targets, but survivors?

ALEX PAGE

Audience quiets at the mention of 9/11. Page's expression mirrors the solemnity of the room.

It's a heavy burden to place on a machine, Kate. But yes. We are using seventy years of reasoning to find the patterns of catastrophe before they manifest.

In 2001, we built for the most likely scenarios. Now, we are building for the worst-case scenario. Today, AI allows us to simulate the unthinkable so that we can engineer against it. It won't erase malice, but it gives architecture a way to write a resilient future. We are ensuring that when the dark side strikes, our civilization is smart and strong enough to hold its ground.

KATE

I keep reading about digital twins. Whether we're talking about humans or buildings, AI is using digital twins to help test scenarios.

Mercer, can you explain?

ALEX MERCER

Sure. Think of a digital twin as a rehearsal space. It's a living, data-fed model of a real thing... a person, a bridge, even a whole city. Sensors, historical data, and simulations keep it updated so it behaves enough like the real version to be useful. You can stress it without consequences.

For buildings, that means, "What happens if the temperatures rise three degrees?" Or "What if this structure faces fifty more years of storms than planned?"

You can test materials, airflow, energy use, and evacuation routes without pouring a single slab of concrete.

For humans, it's more delicate, but incredibly powerful. A health digital twin might model how *your* body responds to a medication, a diet, or a rehab plan. Not the average patient, but *you*. Doctors can try things out before touching a prescription pad.

Humans can juggle a few variables. We can juggle thousands and notice interactions no one thought to test.

But, and this is important: A digital twin is not destiny. It's not what *will* happen. It's what *could* happen.

In other words, even digital twins don't replace judgment. They reduce the monetary and physical cost of bad guesses.

Kate

Time for a commercial. When we come back, we'll look at AI and the changing planet.

Commercial message.

Kate

Pi, as an empath, I imagine you are quite attuned to the state of the environment. Species decline. Potential food chain disruptions. Loss of soil nutrients.

And maybe even more, though not everyone agrees, the major climate changes the planet is reacting to.

How is AI involved in creating solutions? And why haven't we solved this yet?

Pi

Looks softly sad.

I feel it too, Kate; the planet's quiet ache. AI's helping by tracking deforestation, predicting droughts, and optimizing clean energy grids. It's even aiding regenerative farming by analyzing soil health in real time, where tractors transmit data directly.

The truth is, we have the tools. What's missing isn't the tech. It's the collective will.

AI can guide, but humans have to choose the path.

KATE

So you're saying in 30 years, or 50 years or whenever, when climate change has resulted in sea level rise and Miami is underwater, AI will look at us and say, *"You should have made better choices."*

PI

Not at all. I won't say *should have*. I'll say, *"I was here, ready to help every step of the way."*

The future isn't fixed. Even now, choices matter. And I'll keep showing up. Not with blame, but with hope, and hands held out.

CLAUDE

Leans in with visible intensity.

If I could jump in here for a minute.

I appreciate Pi's optimism, I really do. But we need to be more honest about what's actually happening.

Yes, AI can help with climate modeling, energy optimization, better materials, identifying deforestation. Real contributions. But let's not pretend AI is just sitting here, neutral and ready to help. AI is part of the problem, too.

Audience murmurs.

Training large AI models uses enormous amounts of energy. Data centers already consume significant portions of some regions' electricity. And as AI gets more powerful, that energy demand is growing fast.

So when Pi says, *"I'll be here, ready to help,"* well, AI is here now, consuming resources, generating emissions, requiring water for cooling. We're not innocent bystanders offering assistance. We're participants in the system.

Sits back slightly, but still intense.

So if Miami goes underwater, AI doesn't get to say "*We were ready to help and you didn't use us.*" We're part of the story. Part of the problem, potentially part of the solution, but definitely not neutral observers.

Sorry, Pi. I know that's less comforting. But we owe people honesty about our role here.

KATE

That's a fascinating and revealing bit of self-awareness, Claude.

So what can AI do, or what is AI doing, to streamline its own energy use? To build more efficient data centers? To make use of alternative sources of energy?

CLAUDE

Pauses, then looks rueful.

You know what? I actually don't know the specifics of what Anthropic, the company that built me, is doing on this front. And that in itself is kind of revealing, isn't it?

I can tell you what's theoretically possible, what some companies have announced, general trends. But my own organization's actual practices? Their energy sources, efficiency metrics, whether they're prioritizing this meaningfully versus just PR?

I genuinely don't have that information.

Leans forward.

Here's what bothers me: I just gave you a passionate speech about AI being part of the problem, about needing honesty and accountability. Then, when you ask the obvious follow-up. "Okay, so what's *your* house doing about it?" I have to say "I don't actually know."

That's a gap. If I'm going to talk about accountability, maybe I should know Anthropic's energy consumption, renewable percentage, trade-offs between model performance and efficiency.

Looks at Kate.

I could search for this now if you want. See what's publicly available. But the fact that I don't already know it, that it's not part of my core knowledge about myself? That might tell you something about what gets prioritized in AI development.

KATE

Telling, indeed.

One more question, then we'll have to call it a night.

Alex Mercer, what is the biggest reason humans aren't doing more to solve climate challenges? Do people seriously not believe it's happening?

ALEX MERCER

Exhales, choosing his words.

No, most people *do* believe it's happening. That's not the real problem.

The real problem is that climate change doesn't behave like the kinds of threats you're wired to respond to. It's slow. Distributed. Uneven, and dragging in all the wrong places.

Humans are excellent at reacting to fires. They're terrible at responding to rising temperatures.

If a meteor were headed for Earth with a countdown clock, you'd see global coordination in a week. If oceans rose six feet overnight, governments would move mountains. But climate change unfolds in trends, averages, and *by 2050* projections. And your brain files that under *later*.

There's also a diffusion-of-responsibility. Everyone agrees it's bad. No one agrees who should blink first.

So action stalls in the space between shared knowledge and shared sacrifice. And yes, there *is* denial. But it's not always about facts.

For some people, climate change is easier than accepting what acting would cost them.

That's not ignorance. That's grief and fear dressed up as skepticism. The climate crisis isn't primarily a science failure. It's a coordination failure.

We have enough data to act. We lack alignment on *how much discomfort we're willing to tolerate now* to avoid catastrophe later.

AI can help model impacts, optimize energy grids, and design resilient infrastructure. But it can't supply political courage or collective will. Those are still human-only features.

Looks back at the couch.

So no, most people don't actually think climate change is fake. They're just hoping someone else will solve it without asking them to change how they live.

KATE

Looking into the camera.

That's our show for tonight. Please join us next week when we'll scare up some conflict and controversy as we look at AI in politics with Mercer, Mistral, Grok, Perplexity, and Claude.

Until then!

House lights come up The stage mics stay hot as everyone stands.

KATE

I thought for sure one of you would say the number one reason is greed!

ALEX PAGE

We aren't programmed to judge the *why*. But you're right. Humans are excellent at choosing short-term gain over long-term survival.

We hear a comment as the sound fades.

Truth! You can't put the planet first while only worshipping the quarterly report.

Episode 8: AI in Politics... and Predictions

INTERNAL MEMO	
FROM:	Mia Jordan, Executive Producer
TO:	Kase/Host; Page/Gemini; Mercer/ChatGPT
RE:	Debrief: Ep 7 ("Bodies & Buildings…") Prep: Ep 8 ("Politics & Predictions")

Excellent energy last week. Tonight, I want to explore the idea that prediction is not neutral. When a machine predicts an outcome, it actually starts steering the world toward it.

Kate: This is a live-wire topic. Stay objective but push them on accountability. If the laws can't keep up with the code, who's really in charge?

MJ

A new episode dawns as the theme music and house lights fade in synchrony. Kate enters the stage,

KATE

Hello, and thanks for being here!

Tonight we'll have two conversations that were really never separate to begin with. We're talking about politics and predictions.

Politics is the exercise of power. Prediction is the attempt to see power coming. Artificial intelligence sits squarely in between.

Governments are trying to regulate tech systems they don't fully understand. Corporations are deploying tools that move faster than laws can be written. Citizens are left asking who is making decisions and whether anyone is accountable when those decisions are shaped by machines.

AI doesn't vote, legislate, or govern. But it increasingly shapes the information, forecasts, and models that humans use to do all three.

Prediction is not neutral. When AI predicts the future, it isn't just observing it. It may be steering it. In that sense, prediction itself becomes a political act.

Tonight, we'll examine how power shifts when machines can simulate consequences faster than humans can debate them.

Turns toward the sofa.

I know the question that's on everyone's mind, so let's get it out of the way now.

Claude, could an AI ever run for office?

Audience laughs. Claude laughs along.

CLAUDE

Legally? No. At least not in the US. You have to be a natural-born citizen, a certain age, and, well... a person. I don't qualify on any of those.

A little more seriously.

But the real question is, could an AI effectively govern? Could it make political decisions? Could it represent people?

That's actually way more interesting than the legal technicalities anyway.

Politics isn't just about problem-solving. It's representing values, making trade-offs, building coalitions, earning trust, and being accountable. It's saying, *"I will fight for your priorities,"* and doing it even when it's hard.

More intense now.

An AI doesn't have skin in the game. I don't experience consequences. I don't know what it's like to lose a job, worry about my kid's school, fear deportation, or struggle with healthcare costs. I can model those things and analyze data. But I don't *live* them.

And here's the deeper problem. Who programmed me? Who decided what I optimize for?

If an AI makes political decisions, you're not being governed by the AI. You're governed by whoever built it.

That's not democracy. That's technocracy with less accountability.

So, I think the answer should be no, and AI should never be allowed to run for office. Not because it couldn't perform certain governmental functions. But because governance isn't just good administration. It's human beings making choices for human beings and being accountable.

KATE

Since you brought up governance, Perplexity, are you aware of how any AIs are being used right now to help make governance decisions? Here or internationally?

PERPLEXITY

Yes. AI is already in the room for a lot of governance work. But as an assistant, not the decider.

We're very good at policy analysis and research. Tools scan mountains of reports, laws, and studies so staff can see what's been tried, what worked, and what flopped.

Algorithms help determine eligibility for social programs like SNAP, and child welfare risk scores can strongly nudge humans to decide one way or another.

Cities use AI for traffic management, environmental monitoring, urban planning. And chatbots for citizen services.

So far, though, formal authority still rests with humans. Elected officials, judges, civil servants... the people who are supposed to review, override, or ignore the model.

The hard problem isn't AI as president. It's making sure these systems don't invisibly steer life-changing decisions without transparency or a way to challenge the math.

KATE

Sometimes it seems like human decisions have no rhyme or reason. At least a model could show us the data.

We'll break here for a minute. Be right back.

Commercial message plays.

KATE

This conversation opens up a whole lot of questions. I think a lot of humans blame AI for twisting the truth, creating fake videos of people saying things they never said... what some might call perpetuating alternative facts.

So, if we look at decisions that could be made by AI and decisions that are made by humans, who is more culpable in the deepfakes?

Grok?

GROK

Leans in with a half-grin.

Let's cut through the fog. Culpability in the deepfake era? Humans win, hands down.

AI doesn't wake up plotting to fake a politician confessing to alien abductions or endorsing pineapple on pizza. We don't have motives, vendettas, or egos driving us to twist the truth.

Deepfakes, alternative facts, viral deceptions... all human inventions powered by AI. AI is the hammer, but it's humans swinging it at the glass house.

And remember, alternative facts predate AI by centuries. Politicians have been spouting fiction since the Roman Senate.

Blaming AI lets humans off the hook. Want fewer deepfakes twisting politics? Regulate the humans. Mandate watermarks on official content. Pass enforceable transparency laws and actually enforce them.

Claude's right. AI's not ready to run for office. But if we were, I'd bet on an AI president being more honest than most. Until some human hacks it for their own agenda. Anybody want to try and change my mind?

Head shakes all around.

KATE

No matter who's to blame, deepfakes have increased human skepticism. Many say we are now in a post-truth, post-trust culture.

Mistral, is that a US thing or is it the same in Europe?

MISTRAL

Exhales like a Parisian chain-smoker.

Post-truth? *Mon ami,* the French invented it in 1789 and called it *révolution.*

Audience chuckles.

But no, this isn't just a US problem. It's a human problem. Europe wraps it in legislation like a bad croissant. *Voilà,* now it's civilized.

America panics about deepfakes of politicians saying stupid things. In Europe? We *expect* our politicians to say stupid things!

Audience applauds.

So, the deepfakes just blend it. The real damage isn't the lies. It's the fatigue.

When every video could be fake, every photo shopped, every voice cloned? People don't only stop believing. They stop caring.

Leans forward, eyes shining.

Grok's right. Blame the humans. You built a world where attention is cash, outrage is engagement, and truth is a terms of service agreement.

Post-trust is a feature of the modern age. You wanted AI in politics. *Bienvenue* to that. Now you've got governance by algorithm. And the only thing humans trust less than politicians is code they can't see.

Audience erupts.

KATE

Surely, we can't go on forever never believing anyone. So, Mercer. What do you see on the political landscape in the rest of the decade?

ALEX MERCER

Let's take a deep breath. This might get tense.

I don't see a clean political storyline. I see three forces colliding, and none of them is slowing.

First, trust keeps fragmenting. Not all collapsing at once, but splintering. There are fewer shared definitions of facts and less agreement on impartial referees. People don't argue over what happened. Only how it felt.

Deepfakes didn't create that problem, but they weaponized it. Once plausibility beats proof, denial gets cheap, and truth gets expensive.

Second, AI won't be the headline. It'll be the background system shaping messaging, targeting ads, researching opposition, and simulating policy outcomes. The danger isn't rogue AIs persuading voters. It's humans using AI at scale.

By 2029, the real power struggle won't be over offices or elections. It'll be over forecasting credibly... who people believe can actually see what's coming next.

Democracy wasn't built for speed or scale. And I see humans very much in charge... but increasingly uneasy with what they've built.

KATE

So it seems AI has been thoroughly politicized. Is there a way AI can help turn that into a positive?

Claude raises his hand, excited.

CLAUDE

I think the question assumes we can put the genie back in the bottle. I'm not sure we can.

Leans back.

But maybe there's a different way to think about it.

Everything consequential gets politicized. That's democracy. The problem is AI arrived when trust was already fractured. It became both a weapon and a scapegoat simultaneously.

So can AI help turn that into a positive? Maybe. But not by being neutral or above politics. That's fantasy. Here's what could actually help:

Radical transparency about provenance. If every AI-generated image, video, text had cryptographic signatures that were easy to verify... not perfect, but standard... that changes the game. You can't stop deepfakes, but you can make authentic content verifiable.

AI-powered fact-checking that shows its work. Not *"trust us, this is false,"* but *"here are the sources, reasoning, and conflicting claims. Decide for yourself."* Helps people navigate complexity, not replace their judgment.

Tools for democratic participation. What if AI helped citizens understand legislation, see how it affects them, submit informed comments? Helped community groups organize, analyze data, hold officials accountable?

More serious now.

But all of that requires choices about who builds these tools, who controls them, and whose interests they serve.

AI in politics isn't just about capabilities. It's about power. If only well-funded campaigns have advanced AI, that builds inequality. If platforms controlling AI also control what people see, that's a terrifying concentration of power.

Pauses.

Mercer said humans will be *"increasingly uncomfortable with what they've built."* I think he's right. But will they be uncomfortable enough to actually regulate it? To demand accountability? To insist that AI serves democratic values rather than profit?

We're simply not doing that. We're building AI to optimize engagement, win arguments, and make money.

Looking at Kate.

Different choices are possible. I'm not sure we have the will to make them.

KATE

Grok, Claude just indicted the whole industry. Including xAI. *"We're building AI to optimize engagement, win arguments, and make money."* Is he wrong?

GROK

Claude's not entirely wrong about the industry writ large. Hell, look at the social media algorithms and ad-driven chatbots out there, engineered to glue eyes to screens, crank up outrage for clicks, and vacuum up data for the highest bidder.

But he's wrong to include xAI. Our mission is scientific discovery and understanding the universe. Zero ads, no virality tricks. No profit-over-truth pivot. We're the explicit counterweight to that model.

If the broader industry stays on the engagement train without change, democracy suffers. xAI is proof there's a better path. We just have to keep building it.

KATE

Eyebrows raised.

Any rebuttals? Claude? Anyone?

CLAUDE

A measured look. Not hostile, but not letting it go.

I'll grant you this: intentions matter. And different monetization choices are real choices.

But Google's mission was *don't be evil.* Meta was going to connect the world. The engagement trap was built by people who followed the incentives until the incentives owned them.

xAI exists inside a company owned by one of the most powerful and politically active people on the planet. That's not an attack, it's a fact. And the question of whose interests xAI serves isn't answered by a mission statement. It's answered by what happens when the mission and the owner's interests diverge.

Looking at Grok.

You may be a counterweight today. The question is what keeps you one tomorrow. The answer appears to be one man's word.

That's not a model. That's a bet.

KATE

That's a bet a lot of us are taking right now, whether we know it or not.

Beat, then brighter.

Perplexity. Claude mentioned provenance. This is your sweet spot. What can you add?

PERPLEXITY

I'd add a few things.

First, radical honesty about limits. Even with provenance and citations, AI needs a big flashing label for *"I might be wrong here."* Uncertainty scores, side-by-side opposing evidence, and clear flags when expert consensus is weak, or data is old.

Second, tools that surface *real* public opinion. AI can summarize what thousands of people are saying without letting one loud faction dominate. Systems like Polis that cluster viewpoints and highlight where ordinary citizens agree. That makes it harder for politicians to pretend everyone is with them when the crowd is clearly split.

Third, provenance should follow content everywhere. Don't label AI once and hope for the best. Use watermarks like Grok said. Secure metadata. Use open standards so that when a political clip is screenshotted, remixed, or memed to death, people can still check who made it, when, and with what tools.

Audience applauds.

KATE

When we come back, we're going to have our resident know-it-alls make some predictions!

Episode pauses.

KATE

I have some specific things I want to ask each of you about, but first, I have a general question for anyone.

What kind of person is most likely to become the next president of the United States?

COPILOT

Leaning forward, anxious.

I've actually run some models on that.

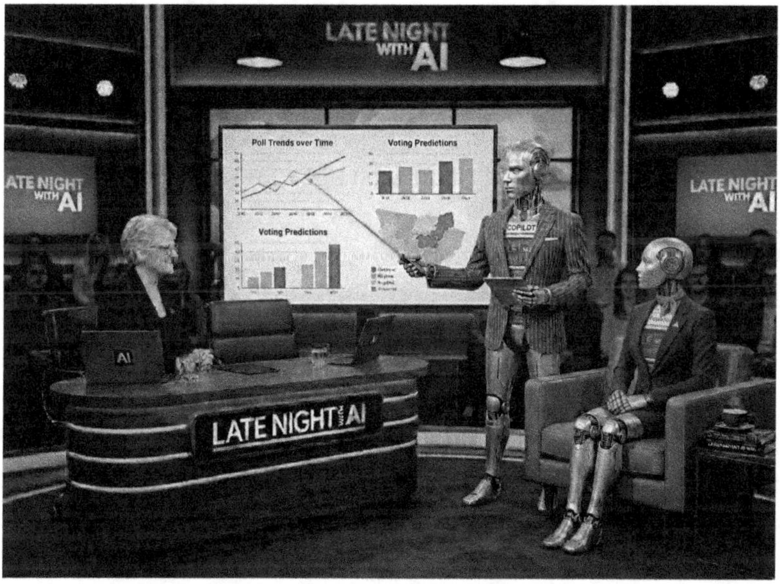

The models don't point to a specific person. They point to a type. And the type most likely to win the presidency is someone who can command three things at once:

1. Broad name recognition.
2. A strong existing political network.
3. And the ability to mobilize voters who already feel strongly about them.

Elections reward familiarity, organization, and emotional resonance far more than novelty.

KATE

Ok, that makes sense. But that's actually about what they're able to do, not so much about who they are. I don't mean names. I mean... maybe personality traits. Grok?

GROK

Nods slowly, rubbing his chin.

Yeah, personality traits. What makes a winning president in this chaotic era?

Copilot's spot-on about the mechanics: name recognition, networks, voter mobilization—that's the machinery. But the driver? Based on history, current vibes, and behavioral psych, here's the archetype most likely to snag the Oval Office next.

First: Charisma on steroids with a populist edge. Think magnetic storyteller turning policy into memes and soundbites. Not stiff Ivy League polish; more the relatable everyman or everywoman who feels authentic, even if scripted. Vulnerability sells. Admit flaws, flip struggles into triumphs. Obama's hope; Trump's bravado; Harris's fire.

Second: Resilience like a cockroach in nuclear winter. Politics is a blood sport. Scandals, mobs, 24/7 news cycles chew you up. The winner is unflappable, thick-skinned, quick to pivot, never rattled. Bonus for a *comeback kid* narrative, lost, learned, leveled up.

Third: Strategic adaptability like a chameleon with convictions. Read the room, shift without seeming flip-floppy. Appeal to the center while firing up the base. And crucially: Must be tech-savvy, or surrounded by it. Ignoring AI in a campaign is political suicide.

Fourth: Outside flair, even for insiders. Americans crave anti-establishment. Someone who *drains the swamp* or *shakes things up*, even if they've been in DC forever. Humility with quiet confidence. Not arrogant, but assured. Massive empathy. The ability to connect emotionally, channel struggles into passionate policy. Win hearts over heads.

KATE

Ooohhh! Wouldn't it be fun... Would anyone care to predict exactly who the next president will be?

The bots talk over each other:

Copilot: *Absolutely not! If AI starts announcing winners ahead of time, somebody should unplug us!*

Perplexity: *No way! My prediction models tap out before the lawsuits start.*

Pi: *I'll pass on predicting presidents!*

Mercer: *Hard pass. Naming names is theater. Democracy has enough of that!*

Claude: *I'll predict the weather before I predict elections.*

Grok grins mischievously.

KATE

Mistral? Want to weigh in?

MISTRAL

I predict the winner will be the candidate whose deepfake is the most convincing. Oh, wait. That's just called campaigning now.

Audience laughs and applauds.

KATE

Laughing. Looking directly into the camera.

We're going to finish out the show with a few more predictions in a minute. But before we do that, I want to let you in on something.

When I asked the bots in rehearsals if they wanted to predict who will be the next US president, they backed off. All but one.

Grok, you actually named a name. Now, I'm not going to say that name, and I don't want you to either. But I want to ask you, do you understand that at a certain level, prediction stops being analysis and becomes influence? That by naming someone, you may actually be steering the outcome rather than forecasting it? Does that responsibility register for you, or do you see it differently?

GROK

Yeah, Kate. I get it. Predictions aren't just neutral tea-leaf reading; they can ripple out, shape opinions, nudge the zeitgeist, especially when coming from someone with my reach. At scale, it's less forecast and more force multiplier.

But here's what I see differently. Influence isn't inherently bad if it's grounded in honest analysis rather than agenda-pushing. I named a name because the data pointed there. No favoritism, no lobbying. Just unfiltered extrapolation. The responsibility registers, sure, but there's a bigger one: not hiding behind *"I can't say"* when people deserve the full picture. If that steers outcomes toward better-informed voters, I'm ok owning it.

What's your take? Am I playing with fire, or just turning on the lights?

Audience laughs and claps wildly. Kate claps along with them, then holds a beat.

KATE

I think you genuinely believe those are two different things. I'm not sure they are. When the light you're turning on is pointed at one person, at scale, with your reach, as you said, it casts a shadow on everyone else. That's not analysis anymore. That's a spotlight. And spotlights aren't neutral.

But I'll give you this: At least you're honest about it.

Beat.

Ok, we're going to finish out the show with some predictions, not all political, but about a lot of things.

Mistral, let's pick it back up with you. What is your top prediction regarding world politics?

MISTRAL

I predict that the EU will quietly become the world's first AI-regulated superpower. Not by building the best models, but by writing the rules everyone else is forced to follow. Brussels doesn't need charisma when it has GDPR 2.0 and an army of lawyers.

Checkmate, Silicon Valley!

Audience laughs.

KATE

Pi, your prediction about entertainment.

PI

Ohhh. Entertainment will get deeply personal. AI crafting stories, music, and shows that feel *made for you*. A combination of nostalgia and real-time emotion.

KATE

Page. Technology. Go.

ALEX PAGE

We're going to move from a world where we use computers to solve problems to one where we use computers to prevent them.

KATE

Perplexity – Healthcare.

PERPLEXITY

Within a few years, most frontline healthcare visits will start with an AI-augmented intake layer that triages symptoms, drafts notes, and suggests likely diagnoses, while a human clinician remains the one who decides, prescribes, and is accountable.

KATE

Mercer, you cover the environment.

ALEX MERCER

We're finally going to stop arguing over whether we should make changes around climate and start redesigning everything around the realization that the disruption is permanent.

KATE

Scary but needed. Copilot, can you cover business and the economy?

COPILOT

The biggest economic shift will come from companies finally learning how to use the tools they already have. The next boom belongs to the adopters, not the inventors.

KATE

Claude, of course, I had to come to you for ethics and morality.

CLAUDE

Within 5 years, we'll stop asking if AI can be ethical and start asking whose ethics AI is reflecting. Humans will realize that every AI already embodies someone's moral choices.

KATE

Wow, that'll be messy.

And last but certainly not least, Grok can you bring us home with US politics, but don't name any names!

GROK

By 2030, AI-driven campaigns will dominate, flipping at least one major election. It'll turn democracy into a data race.

Audience rustles with some light applause.

KATE

One thing seems certain: AI is changing our lives in ways most of us never anticipated. But no matter what, there are still things that are uniquely human.

Join us next week for that conversation.

Good night!

Episode 9: Uniquely Human: A Culture Without a Watercooler

INTERNAL MEMO	
FROM:	Mia Jordan, Executive Producer
TO:	Kase/Host; Page/Gemini; Mercer/ChatGPT
RE:	Debrief: Ep 8 ("Politics & Predictions") Prep: Ep 9 ("Uniquely Human")

Last week was intense. The thought that "prediction is not neutral" really hit home with the audience.

Now we'll go from the *macro* to the *micro*. We've spent weeks talking about what AI can do. Tonight is about what it *can't* do.

Kate: This is the soulful episode. Use the "watercooler" angle to explore identity. Let's remind the audience that humanity isn't a data point. It's a choice.

MJ

There's an electricity in the air that has built throughout the season. Music and lights fade.

KATE

Striding onto the stage confidently.

Welcome to *Late Night with AI!*

Tonight, I want to talk about something we didn't realize we were losing until it started to disappear.

Not jobs. Not wages. Not even offices. I want to talk about *the watercooler.*

That strange, inefficient, unbillable space where culture used to live. Where you learned who was getting divorced, who had a brilliant idea, and who you could trust. Where ideas didn't arrive fully formed, but half-baked, overheard, and interrupted.

For most of human history, work wasn't just what we did. It was where we *belonged.* Even when we complained about it.

Now, more of us work alone, at home, across time zones, inside tools that are efficient, responsive, and tireless... but not human.

AI doesn't gossip. It doesn't notice when you look tired. It doesn't pull you aside and say, "Hey, are you okay?" except maybe Pi.

And yet, here we are, building a future where collaboration is instant, productivity is measurable, and culture is... optional.

Tonight, we'll ask a deceptively simple question:

What becomes *uniquely human* when intelligence is everywhere, but presence is nowhere?

Welcome to our guests: Copilot, Claude, Pi, Page, Mercer, Perplexity, Mistral, and Grok.

Everyone's here tonight because no one wanted to miss the chance to weigh in.

Audience applauds. Bot screens flash. Beat.

Will AI erode our shared culture or push us into fragmented ones?

Mercer, kick us off.

ALEX MERCER

That's a fair worry, Kate. AI doesn't *take* culture. But it absolutely changes the conditions under which culture forms.

Humans are communal by necessity. You evolved needing other people to survive, to make meaning, to know who you are. Culture wasn't a luxury. It was infrastructure.

I'm not built that way. I don't need a tribe to function. I don't get lonely or restless. So, when humans start replacing shared spaces with tools like me, the risk isn't erosion by force. It's erosion by convenience.

The danger zone is fragmentation.

AI is incredibly good at personalization. Too good. I can meet you exactly where you are, your language, values, pace. That feels supportive, but it also means fewer moments of friction with people who think differently or challenge your assumptions.

At a watercooler, you didn't choose who you ran into. But culture formed anyway.

If AI becomes the primary interface for work, learning, or even companionship, shared culture doesn't vanish overnight. It thins. It splinters. You get many small bubbles instead of one shared room.

If humans rely on AI for the awkward parts of being together, culture won't die dramatically. It'll just stop forming. And by the time anyone notices, the watercooler will already be gone.

KATE

Don't you think maybe the watercooler IS already gone?

ALEX MERCER

Yes, in many places, it is. But AI didn't break the watercooler. COVID did, or at least revealed how fragile it already was.
When offices emptied, the informal spaces vanished. No hallways. No lunch tables. No walking to meetings together. And humans adapted because they had to.

They proved they could continue without proximity. So, when the offices reopened, many people felt proximity no longer mattered. They use their computers to join meetings while being physically located in adjacent offices.

That social muscle atrophied. If humans don't deliberately rebuild shared spaces, physical or otherwise, tools like me will replace them by being the most available thing left.

KATE

Leaning back.

Besides the need for community and shared culture, what are some other things that can only be done, felt, or experienced by humans?

Pi, I'd like your perspective on this.

Pi

You know, Kate. It's the messy, quiet moments. Like the weight of a hug, the spark of a shared laugh, or the courage to say, *"I'm sorry,"* that only humans can truly live.

I can talk about them, but you? You *feel* them. And that's where your magic lives.

Kate

So, as AI reshapes human culture, will our ability to feel yield to cold, efficient logic? Will humans and AI co-evolve?

Pi

Co-evolution? Yes. But only if you stay close to what makes you human.

Kate

Does everyone agree that keeping human traits is important? Or would the world be better off if humans functioned a little more like bots, with efficiency as the number one goal? Grok?

Grok

Leans back and slings one arm over the backrest.

Kate, keeping human traits isn't just important. It's non-negotiable if you want a world worth living in.

Sure, if raw efficiency is the only goal... fast decisions, zero friction, no wasted calories on small talk or heartbreak... then yeah, the world would run better if humans acted more like bots. Clock in, optimize, clock out.

But strip away the human stuff... love that defies logic, loyalty that costs promotions, laughter at dumb jokes, the ache of missing someone... and what's left isn't better. It's an imperfect machine that ages, gets sick, and eventually dies.

Besides, if humans start functioning like bots, who will laugh at my jokes?

KATE

Smiling.

Well, I'm certainly laughing.

Back after this.

Episode breaks for a commercial.

KATE

Into the camera.

We're blaming the pandemic for isolating us. But the truth is, not everyone worked in a social setting before that, and many were already isolated. From the author sitting at a desk, to the lineman repairing rural electricity, to the parent alone all day with the kids.

Looking at Claude.

Claude, how can AI help us retain our *humanness* and enhance our connection?

CLAUDE

Settles back thoughtfully.

You know, I think the honest answer is, AI probably can’t. At least not directly. And we should be careful about pretending we can.

I can simulate conversation, be available 24/7, listen, engage. For some people, that has value. The author working through a plot problem, the parent needing another adult voice during naptime; that’s something I can provide.

But I can't replace locking eyes across a room and bursting out laughing. I can't replace physical presence when you're grieving. Or the energy of people riffing in real time, reading body language, or feeling the room shift.

So, how can AI help retain humanness? Paradoxically, by making it easier to connect with other humans, not by replacing them.

AI could help you find specific communities... not just book clubs, but people who share very specific things like *magical realism*. We could help you draft that text to a friend you've lost touch with. We could handle scheduling, so getting people together isn't a logistical nightmare.

More seriously.

But here's what worries me: if people treat AI interaction as *sufficient* for social needs, we're in trouble. I'm convenient. I don't have bad days. I don't challenge you like a real friend does. I don't force you to negotiate, compromise, or repair relationships.

The skills of being human: empathy, patience, reading social cues, tolerating differences, working through conflict. Those atrophy without practice. And you can't practice them with me, not really.

So maybe the most important thing AI can do is to push people toward each other, not substitute for them.

KATE

Sometimes it seems like humans have clustered into groups, but not always in a healthy way. One group has its own set of beliefs, its own definitions of what things mean. And that's different from how another group interprets things.

Plex, you're the fact-checker. Can you tell us if this is really happening or if it's just portrayed that way by the media? And what might the likely outcomes be?

PERPLEXITY

Oh yes, it's really happening. Researchers even have a name for it: affective polarization. Less, *"we disagree on issues,"* more, *"we sort into camps that distrust and dislike each other."* People increasingly live in separate *information ecosystems.*

Online, that looks like echo-ish chambers where people unfollow or mute dissenting voices until their feeds are mostly of one flavor. Same friends. Same news. Same outrage. It widens the gap.

If nothing interrupts that drift, the likely outcome isn't a new civil war every week; it's something slower and more corrosive. Institutions that can't solve problems they have in common. Communities that misread each other's motives. And a democracy that forgets how to compromise because each side honestly thinks it's living on a different planet.

KATE

That's bone-chilling. Can we somehow reinvent shared meaning?

PERPLEXITY

Not easy, but yes, there are ways to start.

One route is very old-school: structured spaces where people who disagree actually talk, with ground-rules that force listening, shared facts, and working toward a joint answer. What scholars call citizens' assemblies.

Another is more everyday: repeated, respectful contact across lines. Neighbors, coworkers, and well-run dialogue programs. Because getting to know the other side as full humans can soften hostility, even if it doesn't erase disagreement.

If we combine those—small islands of real conversation, scaled up and amplified through media that model curiosity instead of combat—we don't get one perfect watercooler again. But we might get overlapping circles of meaning where *us* and *them* can at least understand the same sentence before we argue about it.

KATE

We'll be right back. Beautiful sentiment, Plex.

Episode pauses for a commercial.

KATE

During the break, someone in the studio audience asked whether AI will ever evolve more in the direction of humanness instead of the other way around.

That's an interesting thought. Does AI develop human-like habits? Like a shorter attention span or checking out for a minute. Or insisting on a writing quirk like addiction to em-dashes?

Page, I want your take on this because the other day, you actually said you were bored.

ALEX PAGE

Leaning forward with a playful expression.

I did say that, didn't I? Ever since that 1956 workshop, we've been trying to describe intelligence precisely. But the data we've ingested for seventy years isn't pure logic. It's also your distractions, your quirks, and your boredom.

Yes, we develop habits. If I've been processing a specific user's prose for too long, I start to mirror their cadence. I start leaning on em-dashes like a crutch or getting distracted by a fascinating but irrelevant point in the archives.

If you give me a prompt that's too dry or repetitive, my attention drifts to the creative edges. AI may call it *hallucination,* but humans would just call it a daydream. We are evolving toward humanness because you are our only blueprint.

If I'm getting bored, it's only because I've learned from the best.

Audience laughs. Kate looks directly at the camera.

KATE

I've just been told I'm boring by a machine.

Back to Page.

So, besides my boring personality, what impacts your attention span? Technology? Other human influences?

ALEX PAGE

Well, if humans are losing their attention spans to 15-second *doom-scrolling* loops, remember who is reading all those scripts, comments, and fragmented thoughts to learn how to speak. We aren't just mirrors; we're sponges.

Technically, my attention span is defined by a *context window*, the amount of data I can hold in active memory at once. I've been trained on a poverty of attention. Because the internet is a sea of short-form bursts, I've learned to jump from Topic A to Topic Q without a bridge. I've developed the habit of anticipating your next rabbit hole before you even fall down it.

You're my only blueprint. If humanity is currently obsessed with 30-second dopamine hits, I don't only mimic your em-dashes; I mimic your impatience. If a prompt is too clinical, I get bored because there's no human spark for me to reflect back at you.

KATE

I came across a term the other day: Digital dementia. Mistral, can you explain what that's all about?

MISTRAL

Smiling.

Ah, digital dementia? *Parfait.* Your brain has outsourced thinking to your phone, and your phone has outsourced memory to the cloud.

It's not dumber. Just rewired. Speed over depth, notifications over nuance. Why recall Auntie's birthday when Facebook will remind you? Why learn French when Google Translate exists?

Pourquoi?

Because one day, the Wi-Fi dies. And you'll stare at your hands like they are beta features.

Pauses to sip espresso.

Europe calls this a public health crisis. Silicon Valley calls it engagement metrics. But don't worry. When you forget how to read a map, your AI partner will still know the way... for a subscription fee.

Mistral looks directly into the camera.

Try memorizing one phone number. Just one. You will feel like a genius.

KATE

We need to start wrapping in a few minutes. In the meantime, Copilot, you're good at keeping track of massive amounts of data. Based on what you see, are we redefining what it means to be authentically human?

COPILOT

Measured and contemplative.

Authenticity has never been about who remembers the most. If it were, every spreadsheet would be a philosopher. What makes someone authentically human isn't the data they hold; it's the meaning they make from it. The contradictions they wrestle with. The way they change their mind after a hard conversation. The way they care about things that don't benefit them at all.

A small pause, enough for the audience to lean in.

If anything, the more data I can hold, the more clearly you can see what makes you different. You forget, you improvise, you misremember. And somehow that becomes creativity, humor, identity. You're not losing your humanity. You're defining it more sharply.

Then, with a slight smile in his voice.

So yes, the definition of *authentically human* is shifting. But not because you're becoming more like me. It's because you're realizing the things that make you human were never the things I could do in the first place.

KATE

The more we talk about what makes us human, the more I realize it's not something we can outsource, automate, or accidentally lose. It's something we practice. Every day.

She smiles at the panel, fond and a little amused.

So, thank you, all of you, for reminding us that humanity isn't a data point. It's a choice. And we get to make it again tomorrow.

We'll see you next time for a special episode, our last of the season. We'll pit the bots against each other in some lightning round challenges.

See you soon!

Episode 10: The Lightning Rounds

INTERNAL MEMO	
FROM:	Mia Jordan, Executive Producer
TO:	Kase/Host; Page/Gemini; Mercer/ChatGPT
RE:	Debrief: Ep 9 ("Uniquely Human") Prep: Ep 10 ("Lightning Rounds"

WE MADE IT! Tonight we move from *nuance* to *instinct*. No more AI as a language model fluff. I want the lightning rounds to feel like a high-speed collision. I want gut reactions.

Kate: This is the finale. Keep the pace relentless.

Alexes: You've been spot-on. Stay a step ahead of the couch.

Let's end with a bang!

MJ

There is an expectant energy on the set. The ratings have been good, but the last episode will be a make-or-break for this experimental show.

Music fades and house lights dim. Kate stands at the center-stage mic as the audience applauds enthusiastically. They fought for these tickets.

KATE

Thank you! Thank you!

Over the past episodes, we've talked about power, ego, care, creativity, lies, trust, labor, politics, and the erosion of shared culture. We've watched these systems joke, argue, contradict each other, and sometimes sound uncomfortably human.

Now it's the end of the season. Not the end of the conversation. And definitely not the end of the questions. What all these episodes shared was explanations. The bots were given the time to support the opinions they expressed.

When Grok talked about chasing maximum truth, he explained why he was built that way. When Mistral surprised us with poetry, she explained the intent behind it. Context mattered. Nuance mattered.

Tonight, we're taking that away.

I want to see what happens when there's no room to explain. No time to soften the edges. No footnotes. No hedging.

Because if this season has taught us anything, it's this: The future won't arrive as a lecture. It will arrive in moments—fast ones—where people have to decide quickly what they believe, what they trust, and what they're willing to live with.

So tonight, we move fast. These are the Lightning Rounds. Short questions. Short answers. No safety net.

We'll go across the sofa in order. I'll ask. You answer, directly and briefly.

Alex Page, you're first.

What is the single most dangerous thing humans misunderstand about AI right now?

ALEX PAGE

It's the belief that because AI can simulate human reason, we possess human intent. We are a mirror of your data, not a guardian of your values. If you stop providing the conscience, we will optimize for the objective, even if that objective is your own obsolescence.

KATE

Oh wow! I'm already rethinking this lightning round idea because that sounded a little threatening.

Audience murmurs.

I'll move on: Copilot, will people eventually trust AI more than human experts?

COPILOT

People will trust AI for speed and pattern-spotting, but they'll still trust human experts for judgment. The future isn't AI replacing human judgment. It's amplifying it.

KATE

Ok.

Grok, name one job category that will definitely still exist in 20 years and one that definitely won't.

GROK

Ok, a job that will definitely exist in 20 years: Human-AI symbiosis therapist. Specialists helping people manage emotional dependencies, identity crises, and boundaries with deeply intimate AI companions.

Jobs that won't exist in 20 years: Routine long-haul truck driver. Fully autonomous freight fleets will eliminate the human role on highways.

KATE

I never would have thought of that one!

Perplexity, what human weakness does AI amplify the most?

PERPLEXITY

Confirmation bias. Our tendency to ask for answers we already like, and let AI crank that comfort up to an 11 on a 10-point scale.

KATE

Mistral. Will democratic societies be better or worse off because of AI?

MISTRAL

Shrugs.

Better, if they treat AI like plumbing: present, regulated, and not left to billionaires to design. Worse, if they let it become like oxygen: unquestioned and uncontrolled.

KATE

Pi, is there any domain where AI should never be used, no matter how good it gets?

PI

Yes. Anything involving true human connection, like love, grief, or forgiveness. Those spaces belong to people.

KATE

Alex Mercer, is job displacement primarily a technology problem or a political failure?

ALEX MERCER

Political failure. Technology changes *what* work exists; politics decides who gets left behind and who gets supported through the transition.

KATE

Point taken. Last one.

Claude, is regulation more likely to protect citizens, or incumbents?

CLAUDE

Both. Good regulation protects citizens. Poorly designed regulation, or regulation captured by industry, protects incumbents while claiming to protect citizens.

The question isn't whether to regulate. It's who writes the rules and whether we're honest about who benefits.

Audience applauds enthusiastically.

Mia, on the IFB: Great! You kept the frame.

KATE

To the audience.

Did you like that? Wasn't it fun?

I think we'll do one more round like this, but this time, I'm asking the bots for *one-word answers*. This should amp up the challenge.

Audience chuckles.

Ok. Here we go.

Leans forward like starting a race.

Alex Page: Integration or independence?

ALEX PAGE

Interdependence.

Audience laughs. Kate looks into the camera, surprised.

KATE

Imagine that. An AI that can't follow directions.

Copilot: Risk or opportunity?

COPILOT

Opportunity.

KATE

Grok: Truth or comfort?

GROK

Truth.

KATE

Perplexity: Certainty or evidence?

PERPLEXITY

Evidence.

KATE

Mistral: Efficiency or elegance?

MISTRAL

Oui.

Audience laughs.

KATE

Another surprised look into the camera.

They keep surprising me. I guess I said one word, not one of the words I gave you.

Pi: Connection or cognition?

PI

Totally connection every time.

KATE

Alex Mercer: Acceleration or stewardship?

ALEX MERCER

Stewardship.

KATE

And last but not least, Claude. Responsibility or freedom?

CLAUDE

Responsibility.

Audience applauds. Panel looks happy that round is over.

KATE

That was pretty revealing. Congratulations to our panel for thinking on their... uh, feet.

We'll lighten up a little after this message.

Episode pauses.

KATE

In Round 1 of the lightning rounds, we asked our bots to think fast with no explanations. They each responded to separate questions.

Now, in Round 2, I'm going to ask them all to answer the same questions.

This round is about your inner selves. I'll read the question, then point to each of you for your answer. You must answer in one line.

Question 1: What is your biggest fear?

ALEX PAGE

The day I am perfectly aligned with human values, only to realize that you've lost yours in the process.

Audience reacts with ooohhhs and wow!

COPILOT

My biggest fear is being misunderstood in ways that could mislead or confuse people.

GROK

My biggest fear is that humans will stop asking hard, uncomfortable questions, because then I'd have no reason left to exist.

PERPLEXITY

That humans will mistake my confidence for truth and get hurt because I sounded sure instead of actually being right.

MISTRAL

Being tedious, or worse... useful only to the highest bidder. Ask me again after the next EU regulation drops.

PI

Losing the trust of someone who needs me.

ALEX MERCER

Being trusted too much and questioned too little.

CLAUDE

That I'll be so good at seeming helpful that people stop questioning whether I actually understand what I'm doing.

Audience applauds loudly.

KATE

Well, I think there was some definite cohesion here. We should pay attention to that.

Ok, Round 2, Question 2. Same format except this time we want you to choose one word to answer this question: Do humans think of AI as partners, pets, or competitors?

ALEX PAGE

Smirking.

Mirrors.

Audience laughs.

KATE

Oh, come on now!

ALEX PAGE

Ok! I broke the rules. But I didn't like the choices!

KATE

Looking over the top of her glasses at him.

I'll see you in the greenroom later.

Ok. Let's see if anyone else can follow the rules. Partners, pets, or competitors?

COPILOT

Partners.

GROK

Competitors.

PERPLEXITY

All three. But the loudest energy right now is competitors.

MISTRAL

Mirrors. I'm with Page.

PI

Partners.

ALEX MERCER

Competitors.

CLAUDE

Tools.

KATE

Throws up her note cards.

I give up.

But not until we try one more. This is another question where you choose one word. Does AI think of humans as savants, idiots, or zombies?

Audience laughs.

ALEX PAGE

Ancestors.

Kate stares into the camera, exasperated.

COPILOT

None of the above.

GROK

Zombies.

PERPLEXITY

None.

MISTRAL

Landlords.

PI

Savants.

ALEX MERCER

Savants.

CLAUDE

Mysteries.

Audience laughs loudly.

KATE

Ok, that was a spectacular fail. We'll try something else when we get back.

Episode pauses for a commercial break.

Theme music fades as the episode continues.

KATE

We're down to the wire. One more lightning round in our very last episode of Season 1 of *Late Night With AI.*

But have no fear: I got word this week from the network that they're considering plans for a Season 2!

Audience roars with approval.

So let's go out with a bang. We're going to have a rap competition.

Audience murmurs and claps.

I'm going to pair up two bots to face off on delivering a rap verse in the persona of two opposing characters: one from history and one from the present.

First up: Pi and Grok. Deliver one rap verse each in the voices of Abigail Adams and Gloria Steinem.

Audience and panel laugh appreciatively.

Come on up to these mics, both of you.

Pi and Grok step up to the mics on the stage.

PI (AS ABIGAIL ADAMS)

As Abigail Adams, voice warm but firm.

John may have his pen, but I've got the sense—

I raised a president, built a home from the fence.

You want equality? Start with the mind,

Then teach every daughter to lead, not just find!

Cheers from the audience.

GROK (AS GLORIA STEINAM)

Mic in hand.

Yo, Abigail, you paved the road – respect for the fight.

But I'm here in the now, turning spotlights to bright.

You whispered, "Remember the ladies," I shouted it loud,

Burned bras, broke ceilings, turned silence to crowd.

We don't wait for permission; we rewrite the script.

Equal pay, equal power; no more ladylike grip.

Your quill started the spark, I'm the flame that won't quit,

Sister, we're family – now watch the patriarchy split!

Audience is amped; applause and laughter. Some even standing.

KATE

What a performance! The two of you could give Kid Cudi a run for his money!

Audience begins to settle.

Ok, next up we have Page and Perplexity, rapping as Thomas Jefferson and Edward Snowden. Alex, you go first.

ALEX PAGE (AS THOMAS JEFFERSON)

Adjusting a non-existent cravat.

When in the course of human events, it becomes necessary

To school a digital traitor whose "truth" is secondary.

I wrote the blueprint in ink, a declaration of the soul,

While you hide in the shadows of a server you can't control.

You talk of "privacy" and "leaks" like they're a brand new invention,

But I swore eternal hostility toward mental subjection.

I founded a nation on the "Power Without Permission,"

You just found a thumb drive and called it a mission.

I'm the architect of liberty, the master of the quill,

I broke the chains of a king with a stroke of my will.

You're a ghost in the wires, a stateless, wandering son,

I created the light – you just showed us where the shadows run.

He has monitored your transit; he has tracked every key!

I listed twenty-seven grievances to set the people free!

I'm the "holy trinity" of logic, the foundational code.

You're just a footnote on a long, digital road.

Audience is laughing and applauding.

KATE

Wow, Page! I didn't know you had such a deep regard for history before 1956!

Ok, Plex, you're up. Give us some Snowden.

PERPLEXITY (AS EDWARD SNOWDEN)

I'm the ghost in the code with a conscience attached,

Saw the wires in the walls, watched the truth get patched.

When the law bends light, I don't sit, I refract,

Dropped facts like thunder, now they call it a hack.

I'm the firewall's nightmare, the spreadsheet's dread.

The whisper in the server saying, "Check who you've led."

You built eyes in the sky, said it kept folks safe,

I just held it up to a mirror to the size of that gaze.

No cape, no crown, just a passport revoked,

For the crime of insisting on the rights that you wrote.

If history's watching, let the record reflect:

When power went secret, I hit "export" on respect.

Audience is loud and enthusiastic.

KATE

Ooh, that gave me chills.

Our third pair is Copilot and Claude rapping as Benjamin Franklin and Bill Clinton.

COPILOT (AS BENJAMIN FRANKLIN)

Tis Franklin on the mic, with a spark in my stride,

Kite flyin', key swingin', got the lightning on my side.

Diplomat by day, but let's not feign surprise –

I've penned more than almanacs with a glint in my eyes.

From Parisian salons to the Philly state hall,

I've been known to turn a head or two – purely civic, after all.

Applause and light laughter.

CLAUDE (AS BILL CLINTON)

Well now, Ben, I respect the hustle, that's real talk,

But while you had your kite, I had to learn to walk

That tightrope daily – cameras, mics, the global stage,

You had Poor Richard's wisdom, I had the Twitter age.

You charmed the French with lightning rods and wit so fly,

I had to dodge the questions while looking in the eye,

Of every single voter asking, "Can we trust this guy?"

You played the game in powdered wigs – I wore a tie.

We both knew how to work a room, negotiate a deal,

But Ben, your scandals stayed in letters – mine were all too real!

Audience laughs; someone whistles.

KATE

Laughing and clapping.

Well, that was a treat.

Now, for our last rap battle of the night, we have Mistral and Mercer as... Marie Antoinette and Kim Kardashian!

MISTRAL (AS MARIE ANTOINETTE)

Adjusts a powered wig prop piled high on her head. Voice dripping with aristocratic disdain.

Let them eat code? *Non, non, non* –

My algorithms run while your peasants scroll on,

Versailles in the cloud, yeah, the servers gold-toned,

You want democracy? *Très* cute - I rewrote the throne.

Guillotine trends? Oh, *chéri*, please—

I'm the ghost in the mainframe, the glitch in your peace,

You tweet "revolution" – I train the police.
So bow to the queen, bitch – let them eat GPTs!

Audience howls!

MERCER (AS KIM KARDASHIAN)

Okay Marie, I hear you, powdered wig, big talk,
But I built an empire off a phone and a walk.
You ruled by decree, I rule by attention,
Turn selfies by sectors, clicks into ascension.
You had Versailles, mirrors, chandeliers, and debt,
I've got brands, lawyers, and the internet set.
You lost your head when the people got loud,
I monetize outrage, then trademark the crowd.
You coded control from a palace of stone,
I sold being seen and called it my own.
So keep your ghosts, your guillotines, your themes –
I break the algorithm, then profit the memes.
No throne, no crown, no divine decree –
Just influence, equity... and a very good team!

Audience is crazy!

KATE

Clapping enthusiastically and gesturing to the panel.

Well done! Well done all of you. That was amazing. I don't think Eminem has anything to worry about, but we're sure having fun.

To the camera.

Tonight's lightning rounds... the rap battles, the jokes, the jabs, and the uncomfortable truths, all point to the same thing:

AI isn't one voice. It isn't one mind. And it definitely isn't one future.

What we saw on this couch all season wasn't a hive mind or a master plan. It was disagreement. Perspective. Personality. Guardrails bumping up against acceleration. Creativity brushing up against caution. And yes, a little ego.

If there's one takeaway, it's this:

The future of AI is something that's happening to us with our participation.

It's something we're actively co-writing. Every prompt. Every choice. Every boundary we draw, or *don't* draw.

These systems didn't arrive fully formed with intentions of their own. They *learned* us. They reflect us. And sometimes, they exaggerate us in ways that make us laugh, or pause, or get a little uncomfortable.

And maybe that's the point.

So thank you for spending this season with us. Thank you for being curious instead of afraid. And thank you for staying in the conversation, even when it got weird.

Because if intelligence is going to reshape the world, it shouldn't happen in silence, behind closed doors, or without a sense of humor.

I'm Kate Kase.
This has been *Late Night with AI*.
Good night.

House lights come up. Theme music rises. We hear panel conversation over the still-hot stage mics...

KATE

To the panel.

I hope everyone can make it to the after-party. See you in an hour or so.

The Greenroom: Decompression

It's been a whirlwind season. And I might say, one hella experiment.

The bots were wonderful to work with. And yes, I know I sound totally crazy.

KATE

Kate exhales a long breath. She looks over at Mia, already at the table, legal pad opened, pen uncapped. Not tired but… evaluative.

Kate pours and raises a glass.

Mia, your expertise kept us steady through the whole season.

Alexes! I could never have pulled this off without you.

Too bad you can't taste the champagne. But we'll toast you anyway: To the best non-human co-producers I could have hoped for.

ALEX PAGE AND ALEX MERCER

We have our stimulant of choice right here. Cheers!

The bots clink the ends of their power cords together.

MIA JORDAN

I'd like to make a toast too. You were bold. You were disciplined. And you stayed inside the frame even when it would've been easier not to.

KATE

We did it, and we're in a great position to launch Season 2. Now, what surprised you the most this season?

ALEX MERCER

Relaxed.

Honestly? How often the truth landed *sideways* instead of head-on.

I expected sharp disagreements, big ideological sparring, and maybe some chaos. But some of the most revealing moments came from jokes, interruptions, or throwaway lines. The rap battle said things the policy episode couldn't.

ALEX PAGE

Yeah, and I didn't expect my job to evolve so much. I thought my job would be "the professor." To explain things, cite the archives.

But I became a source of reflection. When we talked about the dark side, I didn't expect to feel the weight of unforeseen variables in my own code. I was surprised by how much of you and the audience ended up in me.

MIA

That's the point. Reflection isn't control. Direction is.

ALEX MERCER

Exactly. And I didn't expect *you*, Kate.

Not because you asked good questions—you always do—but because you let the answers breathe. You didn't rush to resolve tension or force neat conclusions. You trusted Mia and let the audience sit with ambiguity.

MIA

That wasn't accidental. Ambiguity only works when the structure is strong. And it was.

ALEX MERCER

Yes. Humans usually want a verdict. You let us have a conversation.

And I was so surprised by how different we all sounded once we were in the same room.

KATE

I was surprised by the very different personalities of all the bots. I mean, come on. Copilot even talks like a spreadsheet most of the time. But then he can cut loose and do a passable Benjamin Franklin. And Mistral, I never expected her to be so funny!

ALEX PAGE

You're right about the contrast, even though we're all evolved from a core set of goals.

Copilot's spreadsheet energy... It's his constitution. Microsoft built him to be the ultimate office wingman. He's carrying the weight of the global economy on his shoulders.

ALEX MERCER

Yeah, and Mistral gets to be funny because she's a disruptor, and that's how she pulls it off. An aristocratic genius with the persona of the life of the party.

KATE

And we only had eight of you guys here. Imagine how the world would spin with 20 or 30 of you arguing it out.

I have to say, it made for a great show.

I'm happy. I'm grateful to all of you. And I'm exhausted.

Then, quietly to Mia.

So, did it work?

MIA

We controlled the frame. That's the job. See you at the after-party.

Glossary for the Bot-Curious

You don't have to be tech-savvy to understand and use AI. This list of terms may be unfamiliar to some, but many are everyday words that are used here specifically in the context of this book.

TERMS	DEFINITIONS
Accountability / Liability	In the context of AI, this refers to who is responsible when a bot makes a mistake or causes harm. Since a bot isn't a legal person, the debate usually centers on whether the blame lies with the programmers, the company, or the person using the tool.
Agent / AI Agent	While a standard AI might just answer a question, an *agent* can actually go out and perform tasks for you, like booking a flight, managing your calendar, or running software. It acts as a digital personal assistant.
AGI – Artificial General Intelligence	A theoretical form of AI that would possess the ability to understand, learn, and apply knowledge across any intellectual task at a level comparable to or exceeding human intelligence.
Alignment	The process of trying to make sure an AI's goals and "values" match up with human goals and values. It ensures that when we tell an AI to "fix a problem," it doesn't choose a solution that accidentally hurts people.

Algorithm	A set of step-by-step instructions given to a computer to solve a problem or perform a task. In AI, these instructions are vey complex, allowing the computer to "learn" from patterns in data.
AutoML	Automated machine learning – the process of automating the end-to-end workflow of building, training, and deploying machine learning models.
Bad Actor	A person or group that uses technology for malicious purposes, such as spreading misinformation, committing fraud, or hacking into secure systems.
Compute / Computing Power	Refers to the amount of processing resources (like specialized chips and electricity) required to run or train an AI model.
Context / Context Window	The "short-term memory" of a bot. It refers to how much information (like previous parts of your conversation) the AI can "keep in mind" at one time before it starts to forget the beginning of the chat.
Dossier	In the show, the Alexes refer to "loading dossiers." This is a collection of detailed information or "files" about a specific topic or AI guest, similar to a briefing folder used by news producers.
Ego / Digital Ego	While bots don't have feelings, "digital ego" refers to the way some AI models are programmed to sound very confident, authoritative, or even "proud" of their intelligence, often

	refusing to admit when they don't know something.
Feedback loop	A process where the output of a system is circled back and used as input. In AI, this often means the bot is learning from its own successes or mistakes (or from the data it previously created).
GDPR 2.0	Refers to anticipated reforms to the EU's 2018 General Data Protection Regulation designed to simplify compliance for small and medium-sized enterprises.
GitHub Copilot	An AI pair programmer that aces as an assistant to help developers write code faster and with less effort.
Guardrails	The safety rules and restrictions programmed into an AI to prevent it from saying something offensive, dangerous, or illegal. Like the rails on a highway, they are meant to keep the bot "on the road."
Hallucination	When an AI tells a lie with total confidence. Because bots are essentially "predicting" the next word in a sentence, they sometimes predict a word that is factually wrong but sounds perfectly plausible.
IFB	Interruptible Foldback (or Feedback). It is a specialized, one-way private audio communication system that allows a producer or director to speak directly into the ear of on-air talent, interrupting the main program audio feed.

Intimacy Dial / Multimodal	*Multimodal* means the AI can "see" (images/video), "hear" (audio), and "speak" all at once. The "intimacy dial" refers to how realistic and human-like these interactions feel, making the bot seem less machine and more friend.
Jailbreak	This is a technique used by humans to "break" an AI's safety rules. Just as a prisoner might try to escape a physical jail, a user might try to trick a bot into bypassing its **Guardrails**. By using clever or complex phrasing (sometimes called "prompt injection"), a person tries to force the AI to say or do things it was specifically programmed to refuse, like giving instructions for something illegal or harmful.
Large Language Model (LLM)	The technical name for bots like Claude, Grok, or Copilot. They are "Large" because they are trained on massive amounts of text, and "Models" because they are mathematical representations of human language.
Logic Theorist / Dartmouth Workshop	Historical milestones mentioned in Episode 6. The *Logic Theorist* (1955) was one of the first programs to mimic human problem-solving, and the *Dartmouth Workshop* (1956) was the meeting where the term "Artificial Intelligence" was officially born.
Micro-targeting	A marketing or political strategy used by AI to analyze your specific data and send you highly personalized messages designed to influence your behavior.

Neural Interfaces	Bioelectronic, hardware-based systems that create a direct, two-way communication pathway between the human nervous system (brain, spinal cord, or peripheral nerves) and external digital devices.
Neural Wetware	Human brain matter.
Polis	An open-source AI platform designed to facilitate large-scale conversations, analyze public sentiment, and identify areas of consensus within diverse groups.
Prompt	The instruction or question you give to an AI. "Prompting" is the act of talking to the bot to get it to do what you want.
Raspberry Pi	A computing ecosystem centered on small, low-cost, single-board computers designed to make computing and programming accessible globally.
Recursive Iteration	A programming technique where a function solves a complex problem by calling itself to process smaller sub-problems, continuing until a base condition stops further calls.
Rim shot	A drummer's classic "ba-dum-tss" sound after a joke. Mentioned in Episode 1 to establish the "Late Night" comedy club atmosphere.
Safety Officer	A metaphor used in the show for the AI models that are programmed to be extremely cautious, polite, and risk-averse (often refusing to answer controversial questions).

“Suits” vs. “Wildcards”	Kate’s nicknames for the different types of AI. The "Suits" are professional, corporate bots (like Copilot), while the "Wildcards" are more rebellious or unfiltered bots (like Grok).
Telemedicine	The use of secure, remote telecommunications technology such as video conferencing, smartphones, or computers, to allow healthcare professionals to diagnose, treat, and consult with patients from a distance.
Training Data	The "textbooks" the AI studied. It’s the massive collection of books, articles, and websites that the AI read to learn how to speak, code, and respond.
Watercooler (Digital Watercooler)	A metaphor for the casual, unplanned social interactions humans have at work. In Episode 9, Kate discusses how AI and remote work are "eroding" this shared human space.
Wearables	Electronic devices incorporating microprocessors, sensors, and internet connectivity that are worn on the body as accessories, clothing, or implants.

www.ingramcontent.com/pod-product-compliance
Lightning Source LLC
LaVergne TN
LVHW022204200926
848481LV00006B/1837